The Elephant
in the Universe

The Elephant
in the Universe

OUR HUNDRED-YEAR SEARCH
FOR DARK MATTER

Govert Schilling

The Belknap Press of
Harvard University Press

Cambridge, Massachusetts
London, England

2022

FIRST PRINTING

Library of Congress Cataloging-in-Publication Data

Names: Schilling, Govert, author.

Title: The elephant in the universe : our hundred-year search for dark matter / Govert Schilling.

Description: Cambridge, Massachusetts : The Belknap Press of Harvard University Press, 2022. | Includes bibliographical references and index.

Identifiers: LCCN 2021034391 | ISBN 9780674248991 (cloth)

Subjects: LCSH: Dark matter (Astronomy)—History. | Cosmology—History.

Classification: LCC QB791.3 .S32 2022 | DDC 523.1/126—dc23

LC record available at https://lccn.loc.gov/2021034391

Contents

PART III Trunk

The Blind Men and the Elephant

A Hindoo Fable

It was six men of Indostan
To learning much inclined,
Who went to see the Elephant
(Though all of them were blind),
That each by observation
Might satisfy his mind.

The First approached the Elephant,
And happening to fall
Against his broad and sturdy side,
At once began to bawl:
"God bless me!—but the Elephant
Is very like a wall!"

The Second, feeling of the tusk,
Cried: "Ho!—what have we here
So very round and smooth and sharp?
To me 't is mighty clear
This wonder of an Elephant
Is very like a spear!"

The Third approached the animal,
And happening to take
The squirming trunk within his hands,
Thus boldly up and spake:
"I see," quoth he, "the Elephant
Is very like a snake!"

The Fourth reached out his eager hand,
And felt about the knee.
"What most this wondrous beast is like
Is mighty plain," quoth he;
"'T is clear enough the Elephant
Is very like a tree!"

The Fifth, who chanced to touch the ear,
Said: "E'en the blindest man
Can tell what this resembles most;
Deny the fact who can,
This marvel of an Elephant
Is very like a fan!"

The Sixth no sooner had begun
About the beast to grope,
Then, seizing on the swinging tail
That fell within his scope,
"I see," quoth he, "the Elephant
Is very like a rope!"

And so these men of Indostan
Disputed loud and long,
Each in his own opinion
Exceeding stiff and strong,
Though each was partly in the right,
And all were in the wrong!

So, oft in theologic wars
The disputants, I ween,
Rail on in utter ignorance
Of what each other mean,
And prate about an Elephant
Not one of them has seen!

John Godfrey Saxe, 1872

Foreword

AVI LOEB

The term "dark matter" is used to represent most of the matter in the universe—five times more prevalent than ordinary matter, like the atoms that make up stars and planets. But, as the name suggests, we cannot see dark matter. We infer its existence only indirectly through its gravitational influence on visible matter. In this way, dark matter encapsulates our ignorance.

Like all good mysteries, the puzzle of dark matter is enduring. It has intrigued scientists for a hundred years. Observations and scientific theories suggest that dark matter could be made of any number of hypothetical building blocks: weakly interacting massive particles, so-called axions, even atoms that do not interact with ordinary matter or light. Today there is a scientific consensus that dark matter likely came out of the fiery soup during the origins of the universe, an ocean of invisible particles with initially small random motions. Although scientists have not detected any of these invisible particles yet, they have measured the imprint of the fluctuations. Today those dark matter fluctuations are evident in the slightly varied brightness of the cosmic microwave background, the relic radiation left over from the big bang.

Lord Kelvin was the first to offer a dynamical estimate of what we now think of as dark matter. In a talk given in 1884, Kelvin theorized that there might be dark bodies in the Milky Way. Almost fifty years and many ideas later, Swiss-American astronomer Fritz Zwicky estimated that there is more mass in galaxy clusters than is visually observable. In the 1970s evidence for invisible particles was revealed through the pathbreaking work of Vera Rubin, Kent Ford, and Kenneth Freeman. They showed that the dynamics of gas and stars in galaxies imply the existence of invisible mass in a halo that extends well outside the inner region, where ordinary matter concentrates. And in 1983 Moti Milgrom proposed a theory of modified Newtonian dynamics to explain the missing-mass problem. In this alternative hypothesis of gravity, Milgrom postulated that Newton's laws do not apply to galaxies.

Like most explorations in science, historical theories of dark matter found supporters and critics. Milgrom's simple prescription for modified dynamics at low accelerations accounts for the nearly flat rotation curves in many galaxy halos extremely well, even after four decades of scrutiny. But the theory does not adequately account for Zwicky's observed properties of galaxy clusters. Another possibility is that dark matter is strongly self-interacting and avoids galactic cores. And the hypotheses continue.

Throughout this book, Govert Schilling leads us on a captivating tour through the theories of dark matter and efforts to observe it, from early times to the present day. We'll travel with him to astronomical observatories on the ground and in space and to particle detectors in underground caves and tunnels. As we circle the globe, we meet the scientists, the protagonists of the story, who have spent their careers searching for a solution to the puzzle. It is a wide-ranging cast of characters. There are towering figures in the field of dark matter research, like Jim Peebles and Jerry Ostriker. There are younger scientists, true believers, skeptics, and heretics. Through their stories, we garner an extraordinary view into the past, present, and future of one of the deepest enigmas in science.

As *The Elephant in the Universe* shows, the search for dark matter is a work in progress. Hence the abundance of scientific interpretations. But one day all of the pieces of the puzzle will fall into place. It is with Schilling's stellar guidance that we join leading scientists in their crusade to understand this unknown gravitating matter, and, along the way, delight in the mysteries of our universe.

Introduction

In 1995 astronomers announced that they had developed sensitive spectrometers that made it possible to precisely measure the velocities of stars. Within a few years, I reckoned, these tools would be used to discover extrasolar planets: if the spectrometer picked up tiny, periodic perturbations in the velocity of a star, then there might be a massive planet nearby, the gravity of which was disrupting the parent star's movement through space. I decided to start researching a new book on the hunt for exoplanets, in the hope that the breakthrough find could be described in the closing chapter.

In October of that same year, when Michel Mayor and Didier Queloz announced their discovery of 51 Pegasi b—the first confirmed planet beyond our solar system orbiting a Sun-like star—I realized I had to hurry. For most of 1996, I worked on hardly anything else. My (Dutch) book, *Tweeling aarde (Twin Earth),* was published in early 1997. It was one of the first books to cover the initial round of extrasolar-planet discoveries.

Something similar happened about twenty years later. In early 2015 I began researching a book on gravitational waves—tiny undulations of the very fabric of the universe, caused by energetic events like colliding black holes. Albert Einstein's general theory of relativity predicted gravitational waves decades ago, and scientists have been hunting for them ever since. I knew as I started my research that advanced gravitational wave detectors would be going online in a matter of months—new versions of the Laser Interferometer Gravitational-Wave Observatory in the United States and the Virgo detector in Italy. It looked like a discovery couldn't be more than a few years away.

In fact, the first direct observation of gravitational waves came in September 2015 and was announced to the world in February of the following year. Again I put everything aside to complete the book as soon as possible. *Ripples in Spacetime* was published in the summer of 2017.

So when, in early 2018, I started seriously researching a new book on dark matter, I half-jokingly told the astrophysicists and particle physicists I was interviewing that I expected a revolutionary development in the field any day now. Wouldn't it be great if my book were the first to report on the long-awaited solution to the riddle of dark matter? The first to lay out what this mysterious stuff, said to constitute the balance of the cosmos, actually is?

Unfortunately, it didn't happen. So here's the spoiler: when you reach the last page of this book, you still won't know what most of the material universe is made of. But neither do scientists. Despite decades of speculation, searching, studies, and simulations, dark matter remains one of the biggest enigmas of modern science. Still, after reading this book, you *will* have learned a lot about the miraculous universe we live in, and about the ways in which astronomers and physicists have teased out its secrets.

Dark matter challenges our imagination. Like some invisible glue, it is what holds the universe together and what makes it tick. Without it, galaxies would fall apart, galaxy clusters would dissolve, and space would have expanded into oblivion long ago. Dark matter is the most important stuff out there, yet we've only found out about it in recent decades, and no one has a clue as to its true nature.

Well, at least we've learned what it's not, thanks to the work of hundreds of dedicated scientists. Dark matter is not an ocean of ultra-dim dwarf stars. It is not an all-pervading veil of murky gas in intergalactic space. Dark matter is not a population of black holes— at least not the "regular" kind that astronomers are slowly starting to explore. Dark matter isn't even composed of the atoms and molecules that we are familiar with. It's something weird and exotic altogether.

And it has shaped the universe we live in. Dark matter provided the scaffolding for the growth of cosmic structure. It enabled the formation of galaxy clusters, galaxies, stars, planets, and eventually people. However, despite the numerous disciplines and scientists that are involved in studying the problem, we don't seem to be able to really solve it. There have been hints and allegations. Circumstantial evidence and wishful thinking. But so far, not a single convincing detection. No hint of dark matter's true identity.

The story of the search for dark matter goes back to the 1930s, although the mystery wasn't generally acknowledged until some fifty years ago, when astronomers started to wonder about the high rotational velocities of the outer parts of spiral galaxies like our own Milky Way. Before long, particle physicists got involved, as it became evident that the puzzle couldn't be solved without invoking a completely new form of matter. And because of its pivotal role in the evolution of the universe, this new, dark matter also turned into a hot topic in cosmology, the study of the universe at the largest possible scales. Dark matter is a genuinely multidisciplinary area of research that has kept observers, theoreticians, experimentalists, and computer model builders busy for decades.

With so many people working on the problem over such a long time, it's outright impossible to do everyone justice in a book like this. After all, *The Elephant in the Universe* is not a technical book, nor does it pretend to be the definitive history of the field. Instead, this book provides a broad view of dark matter research in all its bewildering variety. Personal stories of many key players give a taste of the ingenuity, perseverance, and—sometimes—stubbornness of scientists who have devoted their professional lives to solving nature's biggest mysteries. I'll take you, the reader, along to remote astronomical observatories and underground laboratories. We'll attend scientific conferences and talk with Nobel laureates and postdoc researchers alike. Unfortunately, due to the COVID-19 pandemic, not all of my planned trips could be realized, and quite a number of interviews had to be conducted on the phone or via Zoom.

Our journey covers a wide range of dark matter–related topics. Although most of the twenty-five chapters could be read as standalone stories, I've arranged them in an order that presents the scope of the mystery and shows how that mystery has evolved. To set the stage, the first chapter introduces physicist James Peebles, who has been called the "father" of the popular cold dark matter (CDM) model and who was the corecipient of the 2019 Nobel Prize in Physics for his contributions to theoretical cosmology. Next, in chapter 2, a visit to the underground Gran Sasso laboratory in Italy gives a preliminary taste of the experimental approach to the dark matter riddle. Dark matter isn't the exclusive province of computer simulations and conference papers. At this very moment, dozens of scientists all over the world are putting theory to the test in hopes of solving this puzzle.

After whetting your appetite by means of this introductory brush with theory and experiment, we travel a century back in time in chapter 3, to learn about the first indications that something was amiss in our understanding of the material contents of the universe. Much later, in the 1970s, physicists realized that galaxies like our own Milky Way cannot be stable without huge, more-or-less spherical halos of dark matter (chapter 4). Pioneers like astronomer Vera Rubin started to realize that the high spin rates of galaxies can only be explained if they contain much more than meets the eye, as described in chapter 5.

Today Rubin's name adorns a brand-new telescope under construction. When completed, it will be one of the most powerful on Earth, an instrument central to scientists' attempts to map the three-dimensional distribution of galaxies in space. That project is an important dimension of dark matter research and the subject of chapter 6. Then, in chapter 7, we delve into the origin of the elements, only to discover why dark matter cannot consist of ordinary atoms and molecules. The decisive role of radio astronomy in proving that dark matter really exists is the topic of chapter 8. This concludes the first part of the book, which is largely focused on astronomical research.

Part II opens with two chapters discussing the growing conviction, in the second half of the 1970s, that the mysterious stuff must be composed of relatively slow-moving ("cold") elementary particles. Such particles fit remarkably well in the theory of supersymmetry—a promising candidate for the long-sought Theory of Everything. Thus dark matter started to play an important role in particle physics, too.

Chapter 11 details computer simulations of the evolution of the large-scale structure of the universe, which seemed to support a candidate for the contents of dark matter: weakly interacting massive particles, or WIMPs. But just as the WIMP hypothesis was emerging, some scientists began to doubt that dark matter is real. Their theory of modified Newtonian dynamics (MOND), discussed in chapter 12, claims that our understanding of gravity needs revision—dark matter hunters may be chasing a chimera after all.

In chapters 13 and 14, we encounter the powerful observational technique of gravitational lensing—the minute deflection of light by the gravity of massive objects. Gravitational lensing was recognized for its potential to rebut the MOND theory and to help scientists find an alternative dark matter candidate known as massive compact halo objects or MACHOs. Alas, the hunt for MACHOs came up all but empty. Instead, another mystery revealed itself in the late 1990s: dark energy. Scientists realized that empty space was expanding at an accelerating rate—a direct result of dark energy. That discovery, and what it might imply for the overall composition of the universe, are the topics of chapters 15 and 16.

Dark energy and the cold dark matter theory have been integrated into a single cosmological model known as lambda-CDM, where the Greek letter lambda (Λ) denotes dark energy. Studies of the cosmic microwave background (sometimes called "the afterglow of creation") provide strong supporting evidence for the model. Moreover, as described in chapter 17, the relic radiation can be compared to the current large-scale structure of the universe to provide a detailed view of cosmic evolution in which dark matter plays an unmistakable role. Even though we still don't know what dark matter is, we've come to realize that it is a key ingredient of cosmology.

Part III concerns current and future searches for dark matter, as well as some of the challenges facing today's cosmologists. In chapters 18 and 19, you'll read about high-tech experiments that seek to detect dark matter particles directly, using ultra-sensitive instruments installed in deep caves and tunnels, which shield the instruments from cosmic rays that would otherwise disturb the measurements. Surprisingly, cosmic rays themselves may carry telltale fingerprints of decaying dark matter particles—the topic of chapter 20.

Chapters 21 and 22 describe a number of worrisome issues that have lately arisen with respect to the lambda-CDM model. As yet, no one knows how serious these problems are, but theorists are already exploring a range of alternative ideas and hypotheses, some of which are presented in chapters 23 and 24. The closing chapter looks ahead, but it is impossible to predict which of the future experiments and observatories will finally solve the century-old mystery of dark matter. Let's just hope it doesn't take another hundred years.

As a science journalist specializing in everything beyond the Earth's atmosphere, I have probably focused a bit more strongly on astronomy than on particle physics, although I've tried to achieve a good balance between the two. I also put more emphasis on past developments, well-established ideas, and current experiments than on new, speculative theories; unconfirmed results; and possible future experiments. If these novelties are here to stay, you will undoubtedly read about them in a future book.

The dark matter hunt continues, but, though it is unfinished, it has already brought us a deeper understanding of a wide range of astronomical and physical phenomena, from fast-spinning galaxies, gravitational lensing, and the large-scale structure of the universe to the birth of atomic nuclei in the big bang and telltale patterns in the afterglow of creation. The search has also spawned other promising theories, fueling speculations about supersymmetry and as-yet-undiscovered denizens of the particle zoo. While searching for the true identity of the bulk of the universe, scientists have unlocked some of nature's most closely held secrets and revealed the stunning complexity of the world we are an integral part of.

PART I

Ear

I

Matter, but Not as We Know It

Phillip James Edwin Peebles, the Albert Einstein professor emeritus of science at Princeton University, fellow of the American Physical Society and the Royal Society, 2019 Nobel laureate in physics, and godfather of the theory of cold dark matter, slowly stands up from his desk and walks toward a bookshelf on the opposite wall, where he picks up two empty plastic bottles.[1]

He blows air over the opening of the larger one. A low, trembling sound fills the room. Next, he puts the smaller bottle to his lips. Another sound, at a much higher pitch. "It's the same principle," Peebles says, with a characteristically meek smile on his face. "Every size has its own favored frequency, and vice versa."

Now wait a minute. Something that simple doesn't earn you a Nobel Prize, right?

Well, it does if you successfully apply it to sound waves in the newborn universe. If you help prove that galaxies can't be stable without loads of mysterious dark matter. And if you then lay the basis for our current standard model of cosmology.

Thus, at 5 a.m. on Tuesday, October 8, 2019, Peebles got the magic phone call from the Swedish Academy of Sciences. He shared the prize with two others but received half of the prize money—totaling some $910,000—"for theoretical discoveries in physical cosmology." "Good God," his wife Alison said when she heard the news. Then Peebles made the daily one-mile walk from his home to his office on the second floor of Jadwin Hall, his eighty-four-year-old head full of jumbled thoughts.

You know, Jim Peebles had never imagined that he would become a cosmologist. Born in 1935 in the Canadian city of Saint Boniface, now part of metropolitan Winnipeg, little Jimmy was a tinkerer—a would-be Gyro Gearloose who studied the pages of *Mechanics Illustrated,* built electrical contraptions, experimented with gun powder, and fell in love with steam locomotives. Oh yes, he'd go out when the northern lights would perform their silent dance in the Manitoba winter sky, and sure, he knew how to find the Pole Star. But astronomy never really captured his tech-savvy mind. When he first learned about cosmology as a graduate student, he found it "exceedingly dull, and ad hoc, and unbelievable," as he once told astronomer Martin Harwit.[2]

That slowly changed after he arrived at Princeton in the fall of 1958. Peebles was a PhD student in the research group of the brilliant physicist Robert Dicke. On Friday evenings, Dicke organized seminars where students, postdocs, and professors freely discussed every scientific topic that piqued their interest. Intimidated at first by other people's grasp of quantum physics or general relativity, Peebles came to cherish these informal meetings, and not only for the occasional beer drinking afterwards. Dicke's preoccupation with cosmology turned out to be contagious.

In 1962 Peebles completed his thesis on the question of whether the strength of the electromagnetic force varies over time. He remained at Princeton as a postdoc, collaborating with Dicke and two other postdocs, David Wilkinson and Peter Roll. In a washed-out 1960s photograph that he showed during his Nobel lecture, Peebles looks tall and slim, with dark, straight hair, spectacles, and an Ice-

David Wilkinson (left), James Peebles (center), and Robert Dicke (right), in the early 1960s, with the receiver they built to study the cosmic microwave background.

landic sweater. There was a lot of distance between graduate school and the black-tie affair in Stockholm.

Peebles's career as a physical cosmologist was launched on a sweltering day in the summer of 1964. In the stuffy attic of the Palmer Physical Laboratory at Princeton, Dicke unfolded his ambitious plans to search for the radiation leftover from the newborn universe—a primordial conflagration millions of degrees hotter than any attic. Scientists expected that radiation from this long-ago event was out there, if only it could be found. Wilkinson and Roll were charged with building the equipment necessary to detect the radiation. "So Jim," Dicke said, "why don't you delve into the theory behind all this?"

Peebles worked out how the hot plasma of the early expanding universe—a mix of electrically charged particles—would have interacted with the energetic radiation to form a dense, viscous fluid, sloshing and vibrating with low-frequency sound waves like a primeval broth. Then, some 380,000 years after the big bang, when temperatures had dropped enough for neutral atoms to form, matter and radiation became "decoupled": no longer did the properties of one command the behavior of the other. And while the radiation could now freely propagate throughout the universe—cooling down to become the faint cosmic background glow that Dicke was after—the matter was left behind with a pattern of over- and underdensities: regions in which the density was just a tiny bit higher or lower than average, with dimensions determined by the frequencies of the original sound waves.

Size relates to frequency, and vice versa, as Peebles playfully demonstrated with his plastic bottles turned musical instruments. The same principle applies to the universe at large, producing that telltale pattern, which physicists call baryon acoustic oscillations. Over time, matter in overdense regions would further condense into galaxies. This is the reason that galaxies exhibit a nonrandom distribution in three-dimensional space: they tend to show up where the early acoustic waves left the densest deposits of matter. In other words, the current large-scale structure of the universe is set by events that took place shortly after the big bang.

It's complicated stuff, and you may forget it for now—we'll get back to baryon acoustic oscillations in chapter 17. Suffice it to say, around his thirtieth birthday, Jim Peebles developed a knack for thinking the grandest possible thoughts—maybe not about life, but certainly about the universe and everything. You don't need to be forty-two for that.

Peebles wasn't even distressed by the fact that radio engineers Arno Penzias and Robert Wilson beat the Princeton group in detecting the cosmic background radiation. From their perch at Bell Laboratories in nearby Holmdel, New Jersey, Penzias and Wilson made the discovery in 1964, just a few months after Dicke convened his team. "Well, boys, I think we've been scooped," a disappointed Dicke told them after he got the phone call about the discovery. But Peebles remembers having felt excited. The discovery meant that he and his colleagues were not engaged in mere speculation—there actually was something out there to be studied.

Peebles had caught the cosmology bug, and he has had it ever since. Soon enough he was lecturing on a topic that once had seemed exceedingly dull and unbelievable. His book *Physical Cosmology* was published in the fall of 1971, a year before he became a full professor.[3] The first edition sits prominently on the bookshelf next to his desk, close to an Albert Einstein action figure.

Physical cosmology. For centuries—no, for millennia—the origin and evolution of the universe as a whole had been treated as something metaphysical. A universe resting on the backs of elephants and giant turtles, a divine act of creation in the not-too-distant past. But finally, the mythological mists began to clear; the sacred stories made way for scientific scrutiny and physical investigation. Cosmology became something you could touch, take apart, understand, marvel at. Even fall in love with—like a steam locomotive.

Fast-forward half a century and Nobel Laureate Phillip James Edwin Peebles, his tall body casually clad in blue jeans and a moss-green sweater, bends over his computer monitor, takes off his glasses to discern the tiny characters on the screen, searches through archived scientific papers, loses himself in historical detail. So much has happened in the past five decades. So many breakthrough discoveries, so many dead-ends. So many riddles! But most of all, the gradual realization that our universe, our very existence, is governed by a mysterious substance. By enigmatic stuff that, for lack of a better understanding, goes by the name of dark matter. To paraphrase *Star Trek,* "It's matter, Jim, but not as we know it."

Yes, there had been early hints, back in the 1930s. But it wasn't until the 1970s and early 1980s that dark matter burst onto the scene, like a surprise protagonist who doesn't appear until the third act and then dramatically changes the plot of the play. There are more things in heaven and Earth, Horatio, than are dreamt of in your philosophy.

The details will have to wait (we still have many pages ahead of us), but there were numerous findings that only made sense in a universe filled with dark matter. Peebles's own research on the clustered distribution of galaxies in space, carried out before astronomers were able to create reliable three-dimensional maps, was suggestive. His theoretical work, together with Princeton colleague Jeremiah Ostriker, seemed to indicate that the stability of disk galaxies was impossible unless they are surrounded by massive halos of dark matter. Not much later, Vera Rubin and Kent Ford of the Carnegie Institution of Washington became the first to convincingly show (or were they?) that the outer parts of galaxies rotate much faster than they would in the absence of dark matter.

And there were ever-more-detailed observations of the cosmic microwave background radiation, the remnant radiation from the newborn universe, revealing that it was as smooth as a baby's skin. It was this unexpected result that led Peebles, in 1982, to propose his cold dark matter model. Here's the problem. Either the hot plasma of the early universe was too smoothly distributed, or the current large-scale structure of the cosmos is too lumpy. You can't have your cake and eat it, too: the feeble force of gravity, acting in an ever-expanding universe, never gets you from the smooth there and then to the lumpy here and now.

Unless.

Unless dark matter is something really weird. A new type of particle, responsive to gravity but not to other fundamental forces of nature like electromagnetism or the strong nuclear force. Not coupled to the early universe's hot radiation bath at all. Moving slowly enough—"cold" enough, in particle physics parlance—to start clustering into what would become an invisible scaffolding, well before the cosmic background radiation was released. A cosmic cobweb of unfamiliar stuff that subsequently pulled in old-fashioned ordinary atoms, which went on to form the luminous galaxies and clusters that we see today. Cold dark matter.

Theoretical discoveries in physical cosmology—that's what half of the 2019 Nobel Prize in Physics was awarded for. Sure enough, in the four decades since Peebles proposed cold dark matter, the theory became hot and illuminating, extremely productive, and an integral part of what is now known as the concordance model of cosmology. (Another key ingredient of this model is dark energy, which is no less mysterious than dark matter and will be discussed in chapter 16.) But Peebles is not the person to brag about it. He feels he has every reason to be modest.

First of all, he says, theoretical discoveries rank second after "real" discoveries. The other half of the 2019 physics Nobel went to astronomers Michel Mayor and Didier Queloz, the pair who in 1995 found the first planet beyond our solar system orbiting a Sun-like star. Now that's a discovery. Or what about the Higgs particle, found in 2012? Gravitational waves, 2015. Those were singular events in which scientists confirmed what was otherwise just (extremely well-informed) speculation. The theory of cold dark matter is nothing like that.

Second, Peebles was, at least for a while, less invested in his theory than were other physicists. Especially when the cold dark matter model was in its younger days, he felt uneasy about cosmologists' enthusiastic embrace. Personally, he didn't take the model that seriously, not back then. "Hey guys, I'm just trying to solve the smoothness problem, and this is the simplest model I can think of that fits the observations. What makes you think this is right? I could make other models, too." In fact, he did; some of those other models didn't need dark matter at all. Granted, the other models didn't stand the test of time. Cold dark matter did.

Third, Peebles recognizes the limitations of his model. We may have this wonderful theory, this concordance model that explains both the properties of the cosmic background radiation and the distribution of galaxies in the universe. But it's full of holes. As Peebles explained it to me, dark matter is a kind of kludge. We are stuck with this ridiculous stuff that we had to dream up and install by hand into our understanding of the universe. We need dark matter, but we don't know what it is. There are just too many open questions.

Which is not to say that we don't know anything about dark matter. Its fingerprints are all over the place; we will encounter them one by one later in the book. And by studying how the mysterious stuff affects its surroundings, we've at least made some progress toward understanding its characteristics.

Still, at times, it seems weird and unbelievable. It's not that surprising to find something new in the heavens, but how could we have missed 85 percent of all the gravitating stuff out there, as dark matter scientists claim? Didn't we just put it in by hand, as Peebles told me, to explain our observations? All those astrophysical fingerprints may constitute convincing evidence, but how much longer are we prepared to wait for the irrefutable detection? How contrived is our solution? How hypothetical our theory?

What if dark matter doesn't exist after all?

I'll confess: every now and then, I find myself in a doubtful mode. Dark matter, dark energy, the enigmatic inflationary birth of the universe, the multiverse for goodness sake—it all seems too far-fetched, too made up. Nature can't be that crazy, and cunning, and cruel, can it? Or is it just my lack of imagination? The inability to accept that nature is not obliged to meet my expectations? Am I like Peter Pan who doesn't want to grow up, and keeps believing in Tinker Bell, in the simple, understandable universe that I learned about when I was a child?

The thing is, I'm not appalled at all by Einstein's general theory of relativity (although I don't fully understand it) or by the existence of neutrinos, to give just two examples. Had I lived in the nineteenth century and heard about relativity and its implications—black holes, gravitational waves, space getting warped and time slowing down—would I have believed any of it without convincing proof? If someone had told me that zillions of uncharged, almost massless particles—that is, neutrinos—pass through my body at light speed every single second, wouldn't I have burst out in laughter? But Einstein's 1915 theory was confirmed four years later, and neutrinos were first detected in 1956, the year I was born. Both belong to the universe I grew up with. The universe I've come to accept. As for the newer and equally counterintuitive fads of nature, maybe I'm just too conservative.

Yet we need to be cautious. Scientists have been wrong before. Frequently, in fact. The road to a better and more complete understanding of the universe is littered with discarded theories and wrong assumptions that have stuck around longer than they should have.

The reason is that scientists are a conservative lot. Even in the face of contradictory evidence, they'd rather tweak an existing theory to accommodate the conflicting data than throw it in the dustbin. Unless an even more successful theory comes along, that is.

For instance, for a long time after seventeenth-century Dutch physicist Christiaan Huygens published his wave theory of light, scientists assumed that "empty" space must be filled by something known as the ether—the medium in which light waves were said to propagate. When later experiments contradicted the first simple ideas about this mysterious substance, physicists didn't discard the concept but adapted it to better agree with the observations. Eventually they had painted themselves into an uncanny corner in which the ether had to be an infinite, transparent, massless, nonviscous, yet incredibly rigid fluid. Only when Einstein's 1905 theory of special relativity rendered the magical ether superfluous did scientists abolish it.

Something similar occurred in the late eighteenth century, when chemists reluctantly had to agree that there was no such thing as phlogiston. This fire-like element was thought to be released by some substances when they combusted. A substance could burn only as long as it could release phlogiston; the fact that fires died when starved of air was understood to mean that a given quantity of air could absorb only so much phlogiston. The attractive idea was promoted around the year 1700 by the German chemist Georg Stahl and had a large following, even when experiments revealed that some metals, like magnesium, became heavier upon burning—a bizarre finding given that, according to the theory, some of its matter was necessarily released. Phlogiston proponents simply concluded that the mysterious stuff must have negative weight! They finally gave in when, in 1783, French chemist Antoine Lavoisier convincingly showed that combustion is a chemical process requiring oxygen, an element whose properties were only then becoming known.

Finally, I can't resist mentioning the best-known example of scientists betting on the wrong horse: Ptolemy's theory of epicycles. Starting from two very plausible assumptions (at least for the ancient Greeks)—namely that Earth is at the center of the universe and that the heavenly bodies move in perfect circles and at constant velocities—Ptolemy came up with his clever geocentric worldview. According to the second-century scholar, a planet moves in a small circle (an epicycle), whose empty center orbits the Earth in a much larger circle, called a deferent.

To comport with the observed motions of the planets in the sky, Ptolemy's model needed a large number of epicycles plus further contrivances, like an arbitrary offset of a deferent's center from the Earth. Nevertheless, the complicated and cumbersome model survived for no fewer than fourteen centuries, until Nicolaus Copernicus and Johannes Kepler finally gave us the current heliocentric world view, in which the planets move at varying speeds along elliptical orbits around the Sun.

So there we are. We've never actually seen dark matter, but we think it must exist. Yet we should always be aware of the silent assumptions that go into our arguments and be concerned about the number of fixes and tweaks that we're allowing ourselves to introduce with no other goal than to keep our theoretical plates spinning. We don't want to be led astray by epicycles again, do we?

It's an unnerving thought. Either there's loads of dark matter out there, frustratingly successful in escaping detection by today's ultrasensitive instruments. Or all these diligent scientists are chasing a phantom.

Jim Peebles isn't confident that we will ever find a final, definitive answer when it comes to dark matter or, indeed, a theory of everything. And even if we arrive at such an all-encompassing description of nature, he says, it's not guaranteed that we'll be able to check it against the real universe. Why should nature give us any evidence at all? Sure, in the past we have been successful in finding the evidence we need to prove and disprove theories, but that could well change in the future. Perhaps we will hit some limit at which the evidence needed is impossible to obtain. It's a thought that worries him every now and then: the horrible possibility that we will end up with a totally internally consistent theory that we cannot test. Alas, there's no guarantee that that won't happen.

No, Peebles is not too discouraged by the fact that he may not live to witness the solution of the dark matter riddle. In his Nobel lecture, he told the audience, "I'm happy to pass on to a younger generation many interesting research questions that I have not been able to solve."[4] Two months earlier, in an interview with Adam Smith, the editor-in-chief of the Nobel Prize website, Peebles expressed hope that this younger generation will be surprised by what is found to be the nature of dark matter. "That is my romantic dream: that we will be surprised yet once again."[5]

At astronomical observatories, in particle physics labs, and in space science institutes all over the planet, many hundreds of young, brilliant scientists are working hard to make Jim Peebles's romantic dream come true. They are not just ready to be surprised, but eager.

It sure looks like dark matter is here to stay. Now we want to know what it is.

2

Underground Phantoms

Junji Naganoma sits at his desk, studying graphs and numbers on his computer monitor. Business as usual, you might think. But this is not your average office. The desk is surrounded by racks, crates, and piles of boxes. Naganoma is wearing a safety helmet and a parka—it's 50° Fahrenheit at most, and there's no daylight. His "office" is a hundred-meter-long cave, dimly lit by floodlights on the damp walls, lined with pipes and cables. Huge pieces of equipment, their function unintelligible just by looking, are arranged here and there. Service tunnels, wide enough for trucks to drive through, connect the cave to two others of similar size. The whole complex sits beneath almost a mile of rock in the Italian Apennines.

Welcome to the Laboratori Nazionali del Gran Sasso, the largest underground physics laboratory in the world.[1] In Hall B, scientists and technicians from twenty-four countries are constructing XENONnT, the latest and most sensitive version of their experiment to directly detect dark matter particles. Naganoma, a postdoc from Japan, is checking test results from a makeshift clean room; the boxes contain dozens of fragile photomultiplier tubes, manufactured by a German university, ready to install. During my visit in late 2019, XENONnT is nearing completion.[2] By the time you're reading this, it is actively collecting data, in pursuit of invisible stuff.

Astronomy has a long history of finding new things that we knew nothing about before. Over time—especially after the invention of the telescope just over four centuries ago—our cosmic inventory has become longer and longer. Astronomers discovered moons orbiting Jupiter, planets in the outer solar system, zillions of stars, interstellar gas clouds, and countless galaxies resembling our own Milky Way. But all these cosmic denizens can be seen, either with a classical "optical" telescope or with instruments that detect X-rays, ultraviolet light, or radio waves—frequencies of light imperceptible to the human eye but which can be discerned by specially designed cameras.

Finding invisible stuff is different. Invisible stuff can be found only if it leaves some kind of mark on its visible surroundings by influencing the properties or behavior of those surroundings. The contents of a sealed cardboard box in my attic are invisible, but I know they're in there because they make the box heavier and harder to move around. A magnet out of view under a table creates tell-tale patterns in iron filings on the tabletop. Griffin, the aptly named protagonist of H. G. Wells's 1897 science fiction novel *The Invisible Man,* leaves footprints in the mud, for all to see.[3] As the saying goes, there's more than meets the eye.

In the wider universe, it's usually gravity that does the influencing—that leaves the mark and thereby suggests to researchers the presence of something invisible. The effects of gravity are fairly easy to distinguish because gravity is unique in the universe. It is the only long-range force in nature that is always attractive. The more mass, the more gravity, the stronger the effect. (In contrast, the electromagnetic force, acting on charged particles, can both be attractive and repulsive, and on larger scales, the effects usually cancel out.) Gravity governs the motions of the planets, the structure of galaxies, and the evolution of the universe as a whole. And, of course, the way apples fall down from trees, as Isaac Newton noted while re-tiring in his family garden some years before he formulated his law of universal gravitation in 1687.

Just by studying the effects of gravity, astronomers got on the trail of the planet Neptune, of the white dwarf companion of the star Sirius, of extrasolar planets, and of the supermassive black hole in the center of our Milky Way galaxy. Like the invisible Griffin, all these objects left their gravitational footprints in the mud, thus be-traying their very existence.

What if you see the marks in the mud but fail to identify the in-visible man? Never mind, you know he must be there, and by closely studying his footprints, you may be able to learn quite a bit about him. Take exoplanets, for example. From their observations, astronomers can deduce a planet's orbital period, the planet's dis-tance from its parent star (and hence, the planet's temperature), and even an indication of the planet's mass. You don't need to actually see the planet; measuring its gravitational influence alone is enough.

At the Gran Sasso laboratory, too, scientists try to learn about in-visible stuff by studying an observable fingerprint. In this case, however, the fingerprint isn't produced by gravity. The cave-bound researchers are looking for dark matter particles, which, if they exist, certainly have mass but cannot be detected on the basis of their grav-itational influence. At the scale of individual particles, gravity is feeble. Gravity's effects only manifest at large scales, when the at-tractive forces of big collections of particles add up. So a single dark matter particle will never exert enough gravitational attraction to become apparent on that basis alone. But since particles, including putative dark matter particles, have mass, such particles will have "oomph." Therefore, it may be possible to spot them individually through their extremely rare collisions with nuclei of "normal" matter—like xenon, the element the scientists at Gran Sasso are using. An interaction between a dark matter particle and a xenon nucleus will produce a tiny flash of light, and it is this flash that scientists hope to detect. Hence the photomultiplier tubes.

Experiments like Gran Sasso's face a bit of a snag, though. The problem is that the same flash of light occurs when an atom is hit by less mysterious subatomic bullets, collectively known as cosmic ray particles. Cosmic rays are energetic messengers from outer space. Most of them are protons—the nuclei of hydrogen atoms. Upon entering Earth's atmosphere, they collide with atoms and molecules of nitrogen and oxygen before reaching the surface of our planet. The result is an "air shower" of secondary particles that reach the surface.

If you're looking for dark matter interactions, these secondary cosmic ray particles are a source of experimental noise. And, as we all know, in a noisy environment it's hard to hear a pin drop. That's where the Apennine limestone comes in. While dark matter easily passes through a mile of rock (after all, the weird stuff rarely interacts with normal matter, or we would have detected it a long time ago), most of the secondary cosmic ray particles—mainly negatively charged muons—are effectively stopped. As far as particle interactions go, the Gran Sasso lab is extremely "silent."

Great. But how do you fund, build, and manage an underground lab as large as a medieval cathedral? Back in 1980, nuclear physicist Antonino Zichichi knew which strings to pull. Italian politicians were contemplating the construction of a highway tunnel beneath the Apennines, providing a fast connection between Rome, on the shores of the Tyrrhenian Sea, and the Adriatic on the east coast. Zichichi, who was president of the Italian Institute of Nuclear Physics (INFN) at the time, suggested excavating just a little bit more. A large subterranean physics laboratory adjacent to the tunnel would establish Italy's leading position in the field.

It all worked out as Zichichi had hoped. The tunnel was completed in 1984, and the INFN lab was established the following year. By 1989 the first underground experiment was running, looking, unfortunately without success, for magnetic monopoles—weird hypothetical particles left over from the big bang. Over the subsequent years, the facility was expanded to a staggering volume of 180,000 cubic meters, its experiments run by some 1,100 scientists from all over the world.

The Gran Sasso Tunnel lies just east of the medieval city of l'Aquila (the Eagle), the capital of Italy's Abruzzo region.[4] Autostrada 24 winds its way from Rome to l'Aquila through an enchanting landscape, traversing so many national parks and nature reserves that it's also known as the Strada dei Parchi. But upon entering l'Aquila, I am painfully reminded that natural beauty comes at a price. The Apennine mountains—the geological spine of Italy—are prone to earthquakes, and large parts of the iconic city center were destroyed by a magnitude 6.3 *terremoto* in the early morning hours of April 6, 2009, killing over 300 people.

l'Aquila is only slowly recovering. The skyline is dominated by building cranes, but many centuries-old churches await full restoration. The steep cobblestone streets are filled with concrete mixers, wheelbarrows, and the sound of clanging and hammering. Pylons and barrier tape are everywhere. Most houses are wrapped in scaffolding and debris netting. It's a depressing sight, and I can hardly imagine the perseverance and determination required to rebuild a city, only to await the inevitable next quake. Suddenly the persistence of particle physicists hunting for dark matter appears futile and luxurious in comparison.

Close to l'Aquila's landmark Fontana Luminosa, an illuminated fountain topped by two bronze female nudes, Auke Pieter Colijn picks me up to drive the remaining ten kilometers or so to the lab's aboveground offices, on the western slope of the Gran Sasso massif. Colijn is the technical coordinator for XENONnT. He is also the one who came up with the experiment's strange moniker. Gran Sasso's previous dark matter experiment used approximately one ton of liquid xenon as a dark matter detector, so it was called XENON1T. But the amount of xenon in the new experiment remained undecided for quite a while, so Colijn suggested the name XENONnT, where n stands for any number. The experiment's xenon load ended up being eight tons, but the nerdy name stuck.

A tall, lean, and easygoing physicist in his late forties, Colijn divides his time between the Dutch National Institute for Subatomic Physics, the Universities of Amsterdam and Utrecht, and Gran Sasso. In Italy most of his colleagues know him simply as AP, because his Dutch name is so hard to pronounce. After Colijn and I make a brief visit to the lab's "external facilities," as they are known—a loose collection of offices and workshops and a canteen that serves terrific espresso—we hit the A24 again, to enter the Traforo del Gran Sasso eastbound. Minutes later, we're beneath 1,400 meters of solid rock, safely shielded from noisy cosmic ray particles. But wait— where's the lab?

It's on the north side of the highway, Colijn tells me, only accessible from the westbound tunnel. He follows an exit to the "nerd roundabout"—the only way back to the tunnel and the entrance of the underground facility. A pretty frustrating detour if you forget your screwdriver, he quips. After passing a security gate and parking the car, we begin our cave walk, wearing sturdy shoes and safety helmets.

So this is where physicists hope to solve the dark matter mystery, I tell myself. If their theories are right, the phantom particles are all around us—it's just a matter of catching them.

It's unexpectedly quiet in the three immense caverns, which are oriented perpendicular to the highway tunnel. On average, two dozen people may be working underground at any given time, but the place is so overwhelmingly large that you hardly notice them. Each dusky hall is about 100 meters long, 20 meters wide, and 18 meters high. Wherever you go, there's a soft hum of equipment and machinery, punctuated by the occasional louder rumbling of huge ventilators and air conditioners.

There is more to Gran Sasso than XENONnT. We work our way around the tanks of the Borexino experiment and stand in awe in front of the Large Volume Detector. These are two giant facilities for the study of neutrinos—elusive, uncharged subatomic particles that may play a key role in solving the mystery of dark matter (see chapter 23).[5] We pass a host of other physics experiments, some modest in size, others as large as a house. Bearing contrived acronymic names like CUPID, VIP, COBRA, and GERDA, they're all devotedly going about their business: a hissing valve here, a vibrating needle dial there, racks of computer equipment and flickering control LEDs everywhere.[6]

With its incomprehensible equipment, eerie atmosphere, and uncanny desolation, the underground lab feels like an abandoned alien freight ship or the postapocalyptic remains of a secret military base. Indeed, what will future archaeologists make of our goals and motives when they hit upon this strange place thousands of years from now?

Finally, we arrive at the site of XENONnT, in Hall B. I have seen the pictures, but that doesn't make the experiment less impressive. Immediately next to a huge cylindrical tank is the eye-catching, rectangular, three-story control building with its futuristic glass walls. A staircase rises on one face of the control building, while the opposite side abuts the tank. The glass structure seems as transparent as the universe is to dark matter itself. Cryogenic equipment, used to keep the liquid xenon at its low temperature of −95°C, is stored on the top level, the control room and data-acquisition systems are on

The XENON experiment, at Gran Sasso National Laboratory, Italy. At left is the giant water tank housing the detector; at right is the control building.

the second level, and xenon storage and purification instruments are on the ground floor—all part of this effort to detect mysterious stuff, the existence of which nobody is really sure about.

The exterior of the ten-meter-tall tank is draped with a giant photo of its interior printed on tarpaulin, giving the impression that the tank, too, is transparent. The tank contains 700,000 liters of water, suspended in which is the detector itself. The detector is another container of sorts, filled with just over eight tons of ultrapure and ultracool liquid xenon. At the top and bottom ends of the container are plates to which are affixed hundreds of sensitive photomultiplier tubes on the lookout for the faint and brief flash of ultraviolet light emitted when a xenon nucleus is hit by a dark matter particle. To increase the chances of detecting the flash, the inner walls of the water tank are clad with Teflon, which has a high UV reflectivity.

Great care must be taken to eliminate all particle interactions that might produce a signal similar to that which scientists expect from the collision of a xenon nucleus and a dark matter particle. Even 1,400 meters of solid rock is insufficient to stop each and every cosmic ray muon; one out of every million penetrates to this great depth. When muons occasionally interact with the surrounding rock, neutrons are produced. These neutrons could easily disturb the experiment because they, too, slam into xenon nuclei to produce flashes of UV light, mimicking the expected behavior of dark matter particles. That's one of the reasons why the instrument is placed inside a large tank of purified water: water is an efficient absorber of neutrons.

Then, there's natural radioactivity, where heavy nuclei gradually decay into lighter ones, emitting alpha particles, electrons, and energetic gamma-ray photons in the process. All of those decay products create background noise in the measurements. The welding seams of the xenon vessel continuously leak radioactive radon atoms. Traces of radioactive krypton are found literally everywhere on our planet since we decided to test and deploy nuclear weapons. And commercially bought xenon always contains minute amounts of radioactive tritium. To minimize the unwanted effects of these contaminants, the liquid xenon is continuously purified by the huge distillation column in the transparent building next to the tank.

The idea for the detection technique (you'll read more about the nitty-gritty details in chapter 18) dates back to the late twentieth century. The XENON project was started in 2001 by Elena Aprile, an Italian-born physicist at Columbia University and "quite a character," according to Colijn. The ever-growing international collaboration has built a succession of larger and larger detectors, from the first three-kilogram prototype to the current eight-ton behemoth, increasing the sensitivity of the experiment with every step. Aprile is still in charge.

Colijn also tells me about XENONnT's big competitor, a similar experiment called LUX-ZEPLIN underway at the Sanford

Underground Research Facility in South Dakota. The leader of the project, Brown University physicist Richard Gaitskell, teamed up with Aprile on XENON for a number of years, but in 2007 the collaboration fell apart. Most of the US groups involved in XENON decided to join Gaitskell in developing their own detector. And then there's PandaX, a large xenon-based dark matter experiment in China's Jinping Underground Laboratory, another contender in the effort to directly detect dark matter.

Despite decades of null results, and despite the alienating atmosphere of the place, the visit to the Laboratori Nazionali del Gran Sasso feels inspiring and energizing. Here and at the few other laboratories like it, brilliant physicists are employing the most sensitive instruments humankind has ever built in order to explore what they believe is the most abundant, yet most mysterious, constituent of the universe. The dedication of these researchers is impressive; their confidence is contagious. Of course we're on the verge of making the breakthrough discovery—if not with XENONnT or its competitors, then probably with one of the other, smaller dark matter experiments at Gran Sasso, with names like DarkSide, CRESST, DAMA, and COSINUS.[7] If only the stubborn particle would choose to reveal itself, however briefly, by leaving a diminutive but detectable mark on our high-tech equipment.

Or might it be a phantom chase after all? Could it be that all our efforts are futile? Are we doomed to fail, either because no detector could ever isolate the elusive particle or because the particle doesn't actually exist? Our day at Gran Sasso is over, and as we walk to the car and then drive out of the tunnel into the sunlight, I ask Colijn for his thoughts on the doom scenario, on the frustration of dark matter physics. What if you've been on a wild goose chase for your whole career?

Surprisingly, Colijn is not flustered by the prospect of failure. For one thing, he's not certain about the existence of dark matter, hasn't chosen a side. "I'll only believe it when I see it," he says. What drives Colijn is not really a desire to discover the dark matter particle. Rather, he is interested in the technical challenge of the experiment itself—the opportunity to help build an incredibly silent instrument, free of every conceivable form of external or internal noise. Constructing detectors like XENONnT will benefit science whatever the results, he says. A new generation of physicists is learning to go to the limit and then push the boundaries from there. His highest reward: the joy of working with a great team.

That evening I have dinner with Colijn and six of his team members, including Junji Naganoma. At a place called Arrosticini Divini on l'Aquila's Via Castello, close to what's left of the medieval Chiesa di Santa Maria Paganica, we enjoy genziana liquor, traditional Abruzzese lamb skewers, and local montepulciano wine. More than ten years after the devastating earthquake, most of the tile-roofed buildings in the tormented city center are still uninhabited, but the bars and restaurants are bustling. The l'Aquilans refuse to give in, determined as they are to overcome even the largest crisis.

Likewise, the young men and women at the table—to me, they seem like boys and girls, really—are determined to face every challenge and overcome any setback on their scientific quest for the answer to one of the greatest riddles nature has yet presented. They weren't even born when astronomers found the first strong evidence for the existence of dark matter, back in the 1970s. Hopefully they will live to celebrate the unfolding of the mystery.

The pioneers of the field weren't that lucky.

3

The Pioneers

Jacobus Cornelius Kapteyn died on June 18, 1922, the same year in which he introduced the notion that dark matter could be a necessary feature of the structure and dynamics of the universe.

Jan Hendrik Oort died on November 5, 1992, sixty years after he was the first to quantitatively determine the amount of dark matter expected to be in the central plane of our Milky Way galaxy.

Fritz Zwicky died on February 8, 1974, forty-one years after he was the first to find evidence for huge amounts of dark matter in a remote swarm of galaxies.

Kapteyn, Oort, and Zwicky were the pioneers of the field. They realized that the universe contains unseen stuff. They thought carefully and deeply about the nature of this puzzle. All three died without witnessing the solution. The old mystery of dark matter plagues us still, like an annoying virus that we have somehow learned to live with.[1]

Mysteries can go away, of course. Today it's hard to imagine how little was known about our universe at the close of the nineteenth century. Astronomers knew of eight planets orbiting the Sun. They had discovered moons, rings, asteroids, and comets, but the origin of the solar system was unknown. Scientists realized that our Sun was just one of many billions of stars, but no one had a clue as to the source of the Sun's energy. Prominent thinkers suggested that impacting meteorites power the Sun, or that the glowing orb was slowly but steadily shrinking, releasing heat as it contracted. It was even thought by some that the Sun burned coal.

Beyond our solar system, astronomy was not much more than stamp collecting. Huge catalogs listed the positions, brightnesses, colors, and sometimes even distances of stars, but little was known about their composition, structure, and evolution—astrophysics didn't yet exist. And although diligent astronomers with ever-larger telescopes had discovered thousands of faint, fuzzy "spiral nebulae," similar to the famous Great Nebula in the constellation Andromeda, no one was sure about the true nature of these celestial objects. Some believed them to be relatively nearby swirling clouds of gas that would one day condense into new stars. Others thought they were humongous collections of stars, many millions of light-years away.

That was the universe in which Jacobus Kapteyn was born, on January 19, 1851, in the small farming village of Barneveld in the Netherlands.[2] The tenth child of fifteen in the home of a stern and devout schoolmaster and his wife, Kapteyn attended his parents' boys-only boarding school before studying mathematics and physics at the University of Utrecht. For a few years, he worked at Leiden Observatory, the world's oldest university observatory. In 1878 he was appointed astronomy professor in Groningen.

Although the University of Groningen did not have its own observatory at the time, Kapteyn still managed to make important contributions to astronomy. Indeed, he acquired worldwide fame by compiling the first ever photographic sky survey. For this, he teamed up with Scottish astronomer David Gill, who used a dedicated 15-centimeter telescope at the Cape Observatory in South Africa to obtain many hundreds of photographic plates of the southern sky. These were shipped to Groningen, where Kapteyn spent five and a half years meticulously measuring the positions of no fewer than 454,875 stellar images, by hand. The resulting *Cape Photographic Durchmusterung* was published in three volumes between 1896 and 1900.

Working on the sky survey ignited Kapteyn's interest in the structure and dynamics of what he called "the sidereal system": How were all these stars arranged in three-dimensional space, and how did they move about? And working with Gill impressed on him the importance and benefits of international cooperation in astronomy, especially for a small country like the Netherlands. Between 1908 and 1914, Kapteyn spent three months each year at Mount Wilson Observatory near Los Angeles, where the director—the famous American astronomer George Ellery Hale—erected Kapteyn Cottage to accommodate Jacobus and his wife Elise during their prolonged visits. (The cottage still exists and is available for rent.)

Those were exciting times, for sure. In 1908 Mount Wilson's 60-inch telescope had just been completed, and local businessman John D. Hooker had granted funds for the construction of a 100-inch instrument, which would see first light in 1917. Mount Wilson was an astronomical Mecca, its huge telescopes destined to unravel the secrets of the Sun, the stars, and the universe.

There certainly was no lack of secrets at the time. For instance, in 1912 Vesto Slipher of Lowell Observatory in Flagstaff, Arizona, discovered that most spiral nebulae were receding from us at improbably high velocities, and no one knew what to make of that. Would the 100-inch telescope be able to finally elucidate the true nature of these strange, fuzzy whirlpools?

Back in Groningen, and later when he returned to Leiden, Kapteyn further developed his own ideas about the universe. From the distribution of stars, he concluded that we live in a more or less lens-shaped conglomeration of almost 50 billion suns, measuring some 45,000 light-years across. And that would be it, according to

Kapteyn. Beyond this collection of glittering lights—our Milky Way galaxy—there would be nothing but empty space. The enigmatic spiral nebulae were just additional inhabitants of this "Kapteyn universe," he firmly believed. And conceivably, there might be other, invisible denizens, too. Dark matter.

Kapteyn was the first to come up with a description of the shape and size of the Milky Way, a description that included a role of dark matter. That moment came in a famous May 1922 paper in *The Astrophysical Journal*. He gave his paper the modest title "First Attempt at a Theory of the Arrangement and Motion of the Sidereal System," but, come to think of it, the endeavor wasn't modest at all.[3] Here was one human being, born an astronomical eyeblink ago on a tiny planet orbiting an inconspicuous run-of-the-mill star, attempting to explain the very structure of everything there is and ever has been. Pretty ambitious.

As for dark matter, Kapteyn, following Lord Kelvin, realized that by mapping out the motions of stars and applying Newton's law of universal gravitation, it would be possible to determine the mass distribution of the "sidereal system."[4] After all, gravity is the great cosmic choreographer, governing the dynamics of the universe. But earlier rough estimates by Kapteyn and by the British astronomer James Jeans suggested that the number of visible stars could not produce the amount of gravity needed to explain the observed stellar motions. As Kapteyn put it in the long abstract of his twenty-six-page article, "It is incidentally suggested that when the theory is perfected it may be possible to determine *the amount of dark matter from its gravitational effect*." Elsewhere in the paper he wrote, "We therefore have the means of estimating the mass of dark matter in the universe."[5]

The means, yes. The precise answer, not yet. Kapteyn was never able to perfect his theory. Six weeks after his monumental paper was published, he died in Amsterdam at the age of seventy-one.

Death always comes too early, but in this case it was particularly sad that the grim reaper wouldn't hold off for another ten years, given the enormously consequential astronomical work of the period. Just sixteen months after Kapteyn passed away, Edwin Hubble (after whom the Hubble Space Telescope is named) discovered that spiral nebulae were in fact "island universes"—that is, galaxies—way beyond our own Milky Way. Six years later, drawing on data provided by Slipher, Milton Humason, and others, Hubble and Belgian cosmologist Georges Lemaître studied the velocities at which these other galaxies were receding from our own and announced that we live in an expanding universe. And in 1932 Kapteyn's student Jan Oort, building on the work of his teacher, concluded that the central plane of our Milky Way galaxy contains large amounts of dark matter. Kapteyn would have loved it all.

In the course of the 1920s, mainly through the efforts of Harlow Shapley, astronomers also discovered that the Milky Way galaxy is much larger and flatter than the Kapteyn universe—more like a Turkish flatbread than a bun—and that the Sun and Earth are located some 25,000 light-years away from its center. Moreover Oort was able to prove, in 1927, that the Milky Way is spinning and that it spins faster near the center and slower toward the outer edge. The lemming-like motions of its constituent stars are orchestrated by the gravity of the system as a whole.

Oort was one of the greatest astronomers of the twentieth century. The father of radio astronomy, he shed light on such diverse topics as galactic rotation, supernova explosions, superclusters of galaxies, and the origin of comets.[6] Oort was born on April 28, 1900, and grew up in Oegstgeest, a village close to Leiden. In 1917 he decided to study physics and astronomy in Groningen, two-hundred kilometers north. It was worth the travel because, as Oort said, "Kapteyn was there." Throughout his long life, Oort expressed great admiration for the man and his work. A brilliant student—and an avid rower and ice skater too, by the way—Oort was particularly intrigued by high-velocity stars, the rare daredevils of the Milky Way that, for some reason, race around where others merely crawl. Dynamics all over again, very much in line with Kapteyn's own research. Eventually, it would be the topic of Oort's 1926 doctoral thesis.[7]

In September 1922, shortly after his mentor's death, Oort moved to Yale University to work with the American astronomer Frank Schlesinger. Then, in 1924, Oort returned to the Netherlands for good. At Leiden Observatory, where he would spend the remainder of his career, he carried out groundbreaking work on the rotational properties of the Milky Way galaxy. This research led to the aforementioned 1932 study. Published in the *Bulletin of the Astronomical Institutes of the Netherlands,* the paper bore the unassuming title, "The Force Exerted by the Stellar System in the Direction Perpendicular to the Galactic Plane and Some Related Problems."[8] It has come to be known simply as the dark matter paper.

It's a tough, thirty-eight-page read, dense with tables, graphs, and equations. Oort basically applies techniques described by Kapteyn ten years earlier and concludes that the central plane of the Milky Way contains quite a lot of invisible mass—something already hinted at by Jeans in 1922 and by Swedish astronomer Bertil Lindblad in 1926.

Oort's inventive approach was to study the motion of stars "up and down" with respect to the central plane of the Milky Way galaxy. From this motion, he could deduce the amount of gravitating matter within the plane. Stars rotate around the center of the galaxy; the Sun, for example, completes a galactic orbit once every 225 million years. But stars also slowly bob up and down, like horses on a carousel. This gives the Milky Way its vertical extent of approximately a thousand light-years. Gravity prevents the majority of stars venturing too far above or below the plane: the bulk of matter—visible and invisible—in the central plane eventually draws drifting stars back again.

By mapping the vertical distribution of stars in the solar neighborhood and measuring their up- and downward velocities, it's possible to calculate the local density of gravitating matter in the galactic plane. Comparing this to the number and estimated mass of visible stars gives you a handle on the amount of dark matter.

The local matter density that Oort arrived at was a mere 0.000,000,000,000,000,000,000,0063 grams per cubic centimeter (6.3×10^{-24} g / cm^3), plus or minus 20 percent. This is an extremely small value—after all, the universe is mostly empty space. Yet it is about three times larger than could be accounted for by stars and interstellar gas clouds. Oort was finding that there was a good deal more mass in the galaxy than met the eye—a sign of considerable quantities of dark matter. Oort also concluded that the dark matter was distributed differently from the visible matter. As he wrote in the paper's summary, "There is an indication that the invisible mass is more strongly concentrated to the galactic plane than that of the visible stars."

Since Oort's study appeared in a Dutch publication—albeit published in English—it took a while before the paper was widely read. Still, Swiss-American astronomer Fritz Zwicky probably knew about it in 1933, when he stumbled upon huge amounts of dark matter in the Coma Cluster of galaxies. Zwicky's publication, in another obscure European journal, appeared one year after Oort's, but the evidence was both more convincing and more alarming. In fact, Zwicky's results were so unsettling that most astronomers chose conveniently to ignore them, in the idle hope that the problem would go away all by itself. For decades, Zwicky's dark matter discovery was the invisible elephant in the cosmological room.

Zwicky was born on February 14, 1898, in Varna, on the Black Sea coast of Bulgaria.[9] His parents, however, were Swiss, and from the age of six on, Fritz lived with his grandparents in Glarus, a village in the eastern Swiss Alps. He studied mathematics and physics at the Swiss Federal Institute of Technology in Zürich, where Albert Einstein had received his teaching diploma in 1900. Zwicky moved to the California Institute of Technology in 1925 to assist Robert Millikan, a giant in the field of solid-state physics, who two years previously had been awarded the Nobel Prize. Before long, however, Zwicky lost interest in solid-state physics and switched to astronomy. Caltech, in Pasadena, was just "down the hill" from Mount Wilson Observatory, with its world-class researchers and telescopes. Soon Zwicky was working with the astronomical hotshots of the time, including Hale, Hubble, and Walter Baade. Brilliant, colorful, outspoken, and iconoclastic, Zwicky would become a hot-shot himself.

Zwicky's 1933 paper made use of a key observational technique in astronomy: measuring redshifts. The redshift is a slight wavelength change perceived in the light that we receive from a rapidly receding light source. The faster an object is moving away from us, the redder it appears. This is akin to the Doppler effect we have all experienced when an ambulance is passing by. Although the siren is producing the same sound the whole time, we hear a higher pitch (a shorter wavelength) as the ambulance is approaching and a lower pitch (a longer wavelength) as it is moving away from us. The perceived wavelength change is proportional to the ambulance's velocity toward or away from us. Light waves behave similarly: if a light source is moving toward us, we perceive a shorter wavelength (a bluer color), while a receding light source appears to be slightly redder.

By the early 1930s, astronomers had measured the redshifts of dozens of galaxies. Surprisingly, these redshifts—and the corresponding recession velocities—had turned out to be larger for more distant galaxies. That remarkable fact had led Lemaître and Hubble to conclude that cosmic distances are not growing because galaxies are racing away from us through intergalactic space but because space itself is expanding, taking the embedded galaxies along for the ride.

Although Zwicky initially hated the idea of an expanding universe, he spent considerable time studying galactic redshifts. In galaxy clusters (huge collections of many hundreds of galaxies grouped together in space), individual cluster members all appear to fly away from us—after all, the distance to the cluster is increasing as a result of cosmic expansion. However, the galaxies in the cluster also move about, like bees in a swarm. The result is that they all have slightly different recession velocities. Some are moving in our direction, to the effect that their recession velocity (and thus their redshift) is a bit lower than the value for the cluster as a whole. Others are moving in the opposite direction, away from us, which slightly increases their recession velocity and corresponding redshift, to values above the cluster average. The observed spread in galaxy redshifts tells you about the motions of the galaxies within the cluster—it is equivalent to a velocity spread. And here, too, these motions are governed by the gravity of the cluster as a whole, just like the motions of stars in our home galaxy are governed by the Milky Way's mass.

Working with observational data others had obtained with the hundred-inch telescope at Mount Wilson, Zwicky estimated the number of galaxies in the Coma Cluster (named after the location in which it is located on the sky). He then assumed that each galaxy

The Coma Cluster of galaxies, where Fritz Zwicky found evidence for the existence of dark matter.

was about a billion times more massive than the Sun and on this basis calculated the total visible mass of the Coma Cluster to be on the order of 1.6×10^{45} grams. In that case, given the spatial extent of the cluster, individual Coma galaxies were expected to show a velocity spread of about 80 kilometers per second.

However, the eight galaxies in the cluster that were bright enough to allow astronomers to measure their redshifts showed a much larger range of velocities, differing from each other by as much as 2,500 kilometers per second. This is far higher than the estimated "escape velocity" of the cluster. In other words, the gravity of

1.6×10^{45} grams of cluster matter is insufficient to keep hold of objects hurtling through space at such tremendous speeds. To prevent the speeding galaxies from flying away into the wider universe, the total mass of the cluster would have to be greater. Much greater.

"In order to obtain [the observed velocity spread], the average density in the Coma system would have to be at least 400 times greater than that derived on the basis of observations of luminous matter," Zwicky wrote. "If this should be verified, it would lead to the surprising result that dark matter"—"dunkle Materie" in the original German text—"exists in much greater density than luminous matter." Zwicky published his elegant but quite unnerving analysis in a Swiss physics magazine called *Helvetica Physica Acta.*[10] The title of the paper translates to "The Redshift of Extragalactic Nebulae," rather underselling the surprising finding within.

Surprising indeed, not to say unbelievable. Jacobus Kapteyn had toyed with the idea that the universe might contain at least some invisible stuff. Fair enough. Jan Oort figured that dark matter in the plane of our Milky way galaxy is about twice as abundant as visible matter. Unexpected maybe, but not completely crazy. But now Fritz Zwicky claimed that the luminous stars and nebulae in the universe constitute no more than 0.25 percent of everything there is. Little wonder, perhaps, that few astronomers paid any attention to the result—it just seemed too bizarre. Also, the whole concept of recession velocities and cosmic expansion was very novel at the time. Surely there was a more satisfying explanation for what Zwicky described as a "noch nicht geklärtes Problem"—a not-yet-solved problem?

Almost ninety years later, the problem of dark matter is still unsolved. In fact, it has become more and more complicated. While Kapteyn, Oort, and Zwicky assumed that dark matter would consist of extremely faint dwarf stars or nonluminous clouds of cold gas, we now realize that it can't be composed of the elementary particles we're familiar with—it's matter, Jim, but not as we know it. And while the initial quantitative results on the invisible stuff were published in small magazines and didn't raise too many eyebrows, the nagging mystery of dark matter is now all over the place, occupying hundreds of astrophysicists, cosmologists, and particle physicists alike.

Kapteyn of course never knew of this development. He died in 1922, in what we now consider the prehistory of cosmology. His ideas about the layout of the universe were revolutionary, but we now know that, for the most part, they were plain wrong.

Zwicky made errors, too, although it took some time for astronomers to realize this. His initial 1933 conclusions about incredible amounts of dark matter in galaxy clusters appeared to be confirmed by redshift observations of thirty galaxies in the Virgo Cluster, carried out by Mount Wilson astronomer Sinclair Smith in 1936. Zwicky's own more detailed study of the Coma Cluster in 1937 also lent ballast to his earlier findings.[11] He summarized these and other results in his 1957 monograph *Morphological Astronomy*.[12] But we now know that Zwicky had underestimated the number of galaxies in the cluster, as well as the average stellar mass of those galaxies. Moreover, his estimate for the distance of the Coma Cluster was much too high, compromising his results.

Still, even after accounting for Zwicky's errors, there remained a discrepancy of a factor of about one hundred between the "visible" mass and the "dynamical" mass of galaxy clusters like Coma. Even the discovery, in the early 1970s, that galaxy clusters contain huge amounts of hot, X-ray emitting gas in the space between their individual member galaxies leaves a mismatch of a factor of ten or so.

So when Zwicky suddenly died of a heart attack in 1974, astronomers were still facing his forty-two-year-old nicht geklärtes Problem.

And what about the third pioneer? After the Second World War, Oort became director of Leiden Observatory and continued to do research in a diverse range of topics. In the late 1950s, he finally returned to his work on the amount of dark matter in the Milky Way's central plane. Using better data, he arrived at more-or-less the same conclusions as he had in 1932. He published these new results in 1960, in another paper in the *Bulletin of the Astronomical Institutes of the Netherlands.*[13]

However, Oort's results didn't stand the test of time. In the late 1980s, Belgian astronomer Koen Kuijken and his thesis advisor Gerry Gilmore of Cambridge University showed that Oort's work suffered from systematic errors, mainly because he had to rely on observations of a certain type of giant stars: the only ones bright enough for spectroscopic velocity measurements at the time.[14] Unfortunately, it is notoriously difficult to estimate the true luminosities, and thus the distances, of these so-called K giants. Moreover, we now know that they're not really representative of the stellar population of the thin galactic disk. Both of these issues affected Oort's conclusions.

Using a novel and very efficient multi-object spectrograph at the 3.9-meter Anglo-Australian Telescope in Coonabarabran, New South Wales, Kuijken and Gilmore observed some 800 more "regular" stars and carried out a much more thorough analysis. In three papers in *Monthly Notices of the Royal Astronomical Society,* they concluded that "available data . . . provide no robust evidence for the existence of any missing mass associated with the galactic disc."[15]

By that time, astronomers realized that our Milky Way galaxy had to be surrounded by an extended, more or less spherical halo of dark matter (we'll come back to that in the next chapter). But apparently, there is no significant excess of dark matter in the central plane of our home galaxy. Oort had been wrong.

Around 1988, Kuijken gave a colloquium about his and Gilmore's research in one of the lecture rooms of Leiden Observatory. Jan Oort, brittle and deaf, was in the audience, his hearing aid plugged almost directly into Kuijken's microphone. He was extremely interested in the new results and later sent a complimentary letter to the young astronomer, who would move to Leiden in 2002 and work as the observatory's scientific director between 2007 and 2012. Even in the last stage of his life, Oort was looking forward to what Kuijken, his contemporaries, and his successors would learn. When I interviewed Oort in 1987, he speculated that "the huge amounts of dark matter that people are finding at large scales in the universe may have to be explained by . . . something completely new. . . . But at the moment, I have no idea where [the solution] might be found."[16]

No one did. In November 1992 Oort died, as old as the century on which he had left so many valuable traces. The Hubble Space Telescope had been launched two years before but still suffered from blurry vision due to a slightly misshapen mirror; astronomers had just obtained the very first detailed satellite measurements of the cosmic background radiation; and particle physicists were toying with the concept of xenon detectors. The golden age of dark matter research was about to start.

Yet, despite the huge advances of the past twenty-five years, today's scientists are still groping in the dark, not so unlike Kapteyn about a century ago, when he first introduced the term "dark matter" in an English-language publication.

When will we finally find the answer to the largest riddle in the universe?

4

The Halo Effect

My husband says dark matter is a reality
not just some theory invented by adolescent computers
he can prove it exists and is everywhere

forming invisible haloes around everything
and somehow because of gravity
holding everything loosely together

The first six lines of "Dark Matter and Dark Energy," written in 2015 by award-winning poet Alicia Suskin Ostriker, neatly sum up the early work of her spouse, theoretical astrophysicist Jeremiah Ostriker. Both are trying to wrap their head around an enigma—Alicia by meticulously sculpting sentences on white paper, Jerry by feverishly jotting down equations on a blackboard. So far, neither approach has solved the mystery. As the ninth line of the poem reads, "[W]e don't know what it is but we know it is real."[1]

Jerry Ostriker is in a hurry. In less than an hour, he has to dash off to a meeting on the birth of black holes. Talk about enigmas! But that's more than enough time to discuss his 1970s work on dark matter halos, right? In his small, orderly office on the tenth floor of Columbia University's Pupin Building, he starts to talk and lecture right away, all the while scribbling down equations on a notepad. Every now and then, he walks over to the blackboard on the wall, chalk in hand, to support or explain his arguments with formulas and crude diagrams.[2]

A short, balding, friendly but intense man in his early eighties, and in a hurry, yes. Ostriker wants to witness, or maybe even find, the solution to the puzzle. In the past couple of years, he's been toying with the novel and speculative concept of fuzzy dark matter (more on that in chapter 24). Crazy perhaps, but so far no one has found a way to disprove it. Maybe there's a 50 percent chance that it's right, he says. No time to explain the details, though. "Read my paper."

It's funny because, back in the 1950s, astronomy wasn't his first choice. Ostriker pursued chemistry and physics. But upon reading a *Fortune* magazine story about the great astrophysicist Subrahmanyan Chandrasekhar, he decided to apply for the PhD program at the University of Chicago, where the famous Indian American scientist worked at the university's Yerkes Observatory, carrying out theoretical research on stellar evolution while editing the prestigious *Astrophysical Journal*.

Chandrasekhar is best known for his work on white dwarfs—ultra-dense stars that pack the mass of the Sun into a volume comparable to the Earth's. A few billion years from now, at the end of its life, our own Sun will collapse into such a weird, compact object, with each cubic centimeter weighing as much as a small SUV. During its final collapse, the Sun will spin up dramatically. The focus of Ostriker's PhD research lay here: the stability of these rapidly rotating white dwarf stars. If spun up fast enough, would they start to lose mass, fly apart, or what? He was still struggling with the stability problem when he moved to the University of Cambridge to work as a postdoc with astrophysicist Donald Lynden-Bell. That was in the mid-1960s; Stephen Hawking was a Cambridge grad student.

Obviously, as is usually the case in astronomy, the stability of a spinning star is not something you can easily test in a laboratory. The nitty-gritty details of the problem also can't be solved purely analytically, with a neat set of equations. Ostriker had to take a numerical approach instead, relying on computer simulations. That may sound easy enough today, but back then computers filled rooms, there were no standard programming languages, and lines of code had to be entered manually by punching holes in paper tape. It took until 1968 before Ostriker got his code to work properly. By that time, he was back in the United States, at Princeton. Working with fellow astrophysicist Peter Bodenheimer and others, Ostriker produced no fewer than eight papers titled "Rapidly Rotating Stars" between 1968 and 1973.[3]

So what's the answer? What happens to a white dwarf—or any other star, for that matter—that spins out of control? We are back in Ostriker's office, where he starts writing down equations again. Angular momentum. Inertia. Viscosity. Potential energy. Pretty complicated if you have to take everything into account. But the outcome is always the same: first the star starts to flatten at the poles, just like the Earth or any other rotating body. But then something peculiar happens. If the rotation rate goes up, the star changes shape. It becomes elongated—no longer an axisymmetric pumpkin but a tumbling dog bone. Eventually the star may even split in two.

I'm not particularly good at equations. What Ostriker describes as "simple physics" is hard for me to grasp. But when he uses plain language, the message comes across. Rotating objects with a lot of angular momentum are happier when they are elongated like a candy bar and tumbling like a majorette baton. He takes a look at his watch. We haven't even started to discuss galaxy halos. But we're almost there. For why would this preference for an elongated shape only work for stars? What about disk-like galaxies like our own Milky Way?

At Princeton, Ostriker had an office in Peyton Hall, just a stone's throw from Jadwin Hall, where Jim Peebles was coming to grips with the cosmic background radiation and cosmology in general. Jim and Jerry got along very well, discussing such diverse topics as primordial nucleosynthesis, pulsars, the large-scale structure of the universe, cosmic rays, and computer programming. Oh, and the stability of spiral galaxies, of course.

Peebles was dabbling in numerical calculations himself, out of interest in the gravitational effects of dark matter in clusters of galaxies. At the time, Princeton didn't have powerful enough computers to handle calculations relevant to that problem, so in 1969 he spent a month at the Los Alamos National Laboratory in New Mexico, where he could use the Department of Energy's number-crunching machines. To make sure he wouldn't disturb the secret programs at what was, after all, a government weapons lab—Peebles was still a Canadian citizen back then—he had to be supervised at all times, usually by a secretary reading a novel.

Simulating gravity in a computer is rather straightforward. You start out with an initial distribution of "test particles," each with a certain mass. Using Newton's laws, you determine the net force that each particle experiences as a result of the gravitational attraction of all the other ones. Next you calculate where each particle ends up after a certain amount of time, as a result of this force. That gives you a new configuration, which serves as the input for the next round of calculations. Larger numbers of test particles and smaller time steps will increase the precision and the credibility of your simulation, but, unfortunately, they will also hugely increase the necessary amount of computer time.

I know all about it. In the early 1980s, I wrote a simple BASIC program for my brand-new eight-bit Commodore 64 home computer. The program would simulate the gravitational chaos resulting from the collision of two rotating disk galaxies—I'm not *that* bad at equations. Each time step took about fifteen minutes to process. After the program ran for a day, I thought the output looked quite impressive, although there probably was little (if any) connection between the pattern of dots on my monitor and the real world. (We'll get back to this sort of modeling, known as high-resolution gravitational N-body simulation, in chapter 11.)

Peebles's experience at Los Alamos piqued Ostriker's interest. What if they could tweak Peebles's code a bit and use it to simulate the evolution of a disk galaxy, to study its long-term stability—or lack thereof? Given that rapidly rotating stars could be expected to deform and split up, it seemed impossible that a flat, spinning disk of billions of stars like our Milky Way galaxy could be stable at all. You'd expect the Turkish flatbread to readily distort into a submarine sandwich, just like a pumpkin-shaped star turns into a dog bone if you spin it up fast enough.

Sure enough, the very first two-dimensional numerical simulations of rotating disk galaxies, published by astronomers Richard Miller, Kevin Prendergast, and Bill Quirk in 1970 and by Frank Hohl in 1971, showed just that: the initially circular disk turns into an elongated, bar-like structure, and the galaxy's stars end up in wildly elliptical orbits—very different from the orderly circular motions observed in the Milky Way.[4] With the help of Princeton's Ed Groth, Peebles and Ostriker developed a program that would run on the university's computer, while adding a third dimension to the simulations. Their results agreed very well with those of Miller, Prendergast, Quirk, and Hohl. As Ostriker and Peebles wrote in *The*

Astrophysical Journal, "Axisymmetric, flat galaxies are grossly and ir-reversibly unstable."[5]

But their now-famous December 1973 paper went much further. It was one thing to show that orderly rotating disk galaxies are unstable; it was quite something else to explain why we still see them all around us in the universe. What enables our Milky Way to keep up its orderly appearance? What prevents it from flying apart?

Expectantly, Ostriker looks up from the notepad he's been scribbling on, as if I have to provide the answer. It's just simple, intuitive physics, he says—everyone could have thought of this. Spinning, low-mass galaxies are unstable; more mass would help. But if that additional mass is also located in the rotating disk, the galaxy would be just as unstable as before—after all, the simulations showed that it's the disk shape itself that leads to instability. No, the extra mass needs to be distributed in a huge, more or less spherical halo, not taking part in the orderly rotation of the disk.

Intuition comes first; mathematics follows suit. New computer simulations, using the same code but quite a different initial distribution of test particles, confirmed the hunch: if there's a lot of gravitating mass in a spherical halo (maybe up to two and a half times the mass in the disk), the flat, rotating galaxy remains stable and keeps its regular appearance. As Ostriker and Peebles wrote in their paper, "A massive halo seems the most likely solution for our own Galaxy." And, of course, for other "cold"—that is, orderly rotating—disk galaxies, too.

The landmark publication, "A Numerical Study of the Stability of Flattened Galaxies: or, can Cold Galaxies Survive?" is mentioned in every anthology of dark matter research. Ostriker and Peebles, you'll read, were the first to convincingly show that galaxies like our own Milky Way can't be stable without huge, massive halos of

Artist's impression of the invisible halo of dark matter (pictured as a diffuse cloud)
surrounding a Milky Way–like spiral galaxy.

dark matter. (Later research has revealed that large, random stellar
motions in the cores of galaxies can also stabilize flat, rotating disks,
but most astronomers believe that the original hunch was right
anyway.) However, the phrase "dark matter" doesn't appear a single
time in the fourteen-page paper. If scientists have come to view
halos as fonts of mysterious dark matter, Ostriker and Peebles were
not willing to go that far in 1973. True, it was evident that the mass
in the halo couldn't emit a lot of light—after all, spiral galaxies are
not observed to be embedded in luminous spheres. But who knows,
large numbers of very faint stars might do the trick.

   In fact, astronomers already knew of galactic halos—a term first
coined in the 1920s—and also knew that these halos contained stellar
inhabitants. For instance, dozens of so-called globular star clusters,
each containing up to a few hundred thousand individual stars,
swarm around the center of the Milky Way in a roughly spherical

distribution, with a strong concentration toward the galaxy's core. So as far as Ostriker and Peebles were concerned, there wasn't any obvious reason why the halo couldn't be home to countless dim dwarf stars, too, increasing the halo's mass enough to stabilize the Milky Way. As Jan Oort had written back in 1965, "Some 5 percent of the total mass of the galaxy may be estimated to consist of [orange and red dwarf stars]. There is no way for estimating how much more mass there may be in the form of intrinsically still fainter stars. The real mass of the halo remains entirely unknown."[6]

So how massive are galaxy halos? In other words, how massive are spiral galaxies? That was the topic of a second and much more concise *Astrophysical Journal* paper that Ostriker and Peebles wrote a year after their first, in 1974, with Israeli astrophysicist Amos Yahil, a visiting researcher at Princeton at the time.[7] "In fact, it's the more relevant of the two papers," Ostriker says. However, the black hole meeting, elsewhere in the Pupin Building, is about to start in fifteen minutes or so, and there's not much time left to discuss it in detail. "Just read the paper," he urges.

It's a bold publication, with a bold title—"The Size and Mass of Galaxies, and the Mass of the Universe"—and quite a number of bold statements. The very first line may even have come as a shock to some readers back in 1974. "There are reasons, increasing in number and quality," the authors wrote, "to believe that the masses of ordinary galaxies may have been underestimated by a factor of 10 or more." In just four pages, Ostriker, Peebles, and Yahil summed up and reviewed the various indications that skinny-looking spiral galaxies may in fact be obese heavyweights. Much more massive than you would guess on the basis of their looks.

You can't put a galaxy on a scale, but there are other ways to estimate their mass. Just look at how strongly they tug on their neighbors. Our Milky Way galaxy is surrounded by dwarf galaxies. The dimensions—and relatively sharp edges—of these satellites are governed by the interplay between their own internal gravity and the Milky Way's mass. Elsewhere, the dynamics of small groups of galaxies and of galaxy pairs, orbiting each other, provide information on galaxy masses. And wherever you look, you see the same thing: evidence for much more mass than you would expect on the basis of the amount of light you're seeing. Or, in the language of astrophysicists, a very high mass-to-light ratio.

Talking about tugging: our Milky Way and the neighboring Andromeda galaxy provide another neat argument for huge galaxy masses. Despite the overall expansion of the universe, the two spirals are now approaching each other at a relative velocity of 110 kilometers per second, responding to their mutual gravity. Back in 1959, Franz Kahn of the University of Manchester and Leiden astrophysicist (and former Oort student) Lodewijk Woltjer concluded that the high approach velocity could only be explained if the total mass of the two galaxies and everything in between them were on the order of a trillion solar masses—again, a very high mass-to-light ratio.[8]

At a smaller scale, there were the brand-new radio astronomy results (described in more detail in chapter 8), which seemed to suggest that spiral galaxies also have a high mass-to-light ratio. These somewhat preliminary findings appeared to indicate that the outermost regions of spirals rotate unexpectedly fast, which suggests that the galaxies contain lots of mass. If they did not, they would break apart at such high speed. Yet the visible light output of a galaxy strongly declines at a certain distance from the center. So here, too, the amount of emitted light is out of alignment with the amount of mass that must be present.

A stronger pull, a larger extent, a higher mass. It looked like astronomers had indeed severely underestimated the sheer significance of galaxies—the gravity of the matter, so to say. And where would all this low-luminosity matter hide? Right: in the halo, which Ostriker and Peebles had shown to be necessary anyway, to explain the system's stability in the first place. In their 1974 paper with Yahil, they still suggested that a spiral galaxy's halo could mainly consist of faint stars (this second paper also doesn't mention dark matter at all), but by now, it all started to feel a bit uneasy. A ten-fold increase in mass—could there really be that many faint dwarf stars?

Moreover, there was the second part of their paper's title: the mass of the universe. If you know the average mass-to-light ratio of galaxies, and if you estimate the number of visible galaxies out to a certain distance, it's pretty straightforward to calculate the average mass density of the local universe. (Even I could do that.) The answer that Ostriker, Peebles, and Yahil arrived at was 2×10^{-30} grams per cubic centimeter, or about one hydrogen atom per cubic meter if you would smear out all the mass in all the galaxies evenly throughout space. Writing in *Nature,* three Estonian astronomers, Jaan Einasto, Ants Kaasik, and Enn Saar independently reached a similar conclusion.[9]

But this number, incredibly small though it is, seemed impossibly large. In the early 1970s, cosmologists and nuclear physicists were starting to understand the synthesis of chemical elements during the big bang, and comparing these results with the observed amount of deuterium (heavy hydrogen) in the universe told them that the current mass density of the universe was much lower. (More on this in chapter 7.) In other words, it looked like there just weren't enough atoms in the universe to explain the large galaxy masses that the Princeton team and the Estonians arrived at.

Matter, but not as we know it.

Ostriker needs to go. He gives me a copy of *Heart of Darkness,* the 2013 book he wrote with the British astronomer and science writer Simon Mitton.[10] In the elevator Ostriker tells me about a talk he gave at the 1976 meeting of the National Academy of Sciences in Washington, DC, describing his work with Peebles and Yahil. "Much later, someone asked me why I hadn't mentioned the work of Vera Rubin in that talk," he says. I nod understandingly—wasn't she the first to establish that the outer parts of galaxies rotate too fast? "Vera was a great astronomer," Ostriker continues, "but at that time, she only had very preliminary results. The paper that brought her well-deserved fame wasn't published until 1980."

I leave the Columbia campus somewhat confused. Unfortunately, I can't talk to Vera Rubin anymore; she died in 2016. But her collaborator, Kent Ford, should still be around somewhere—what's *his* story? In a Starbucks coffeehouse across Broadway, I check my email and organize my notes. So much happened in the 1970s, almost half a century ago. So many surprising results, all pointing in the same direction: our expanding universe is governed by dark, mysterious stuff that may not even resemble the matter that stars, planets, and people are made of.

Outside, in the January cold, small groups of students, young parents with children, hurried businesspeople, and an endless stream of cars and cabs pass by. We're all busy living our lives as best as we can, usually unaware of our place in the Milky Way galaxy, let alone of the huge, dark envelope it sits in. Completely unaware of the fact that, without this mysterious substance, we probably wouldn't be here.

So important, and still we don't know what it is.

I look up the last lines of the poem "Dark Matter and Dark Energy," which Alicia Ostriker wrote in the year her husband was a corecipient of the prestigious Gruber Prize in Cosmology. Beautiful as the verses are, they too don't offer any answers.

the way every human and every atom
rushes through space wrapped in its invisible
halo, this big shadow—that's dark dark matter

sweetheart, while the galaxies
in the wealth of their ferocious protective bubbles
stare at each other

unable to cease
proudly
receding

5

Flattening the Curve

W. Kent Ford Jr. has a beetle named after him.

Pseudanophthalmus fordi was discovered in two of the many karst caves of rural Virginia by Tom Malabad of the Virginia Division of Natural Heritage. Since both the Russell's Reserve Cave and Witheros Cave are located on Ford's property, the new species was named after the retired astronomer.

A plaque featuring the naming citation and a photo of the rare beetle is among the exhibits Ford has prepared prior to my visit. On the coffee table in front of the friendly, stocky, and balding eighty-eight-year-old is a stack of books and papers. Large, mounted prints of black-and-white photographs are displayed against the wall, on the dresser, and on the sofa.[1]

"Here's Vera at the plate-measuring machine at DTM," he says, referring to the Department of Terrestrial Magnetism at the Carnegie Institution of Washington. "This is her at the telescope at Kitt Peak. Here's a close-up of my image tube. This one is much later: we're hugging each other upon meeting at a Carnegie colloquium."

The centerpiece of his visual trip down memory lane is the famous plot of the rotation of the Andromeda galaxy. Together with Vera Rubin, Ford showed that the outer parts of Andromeda rotate

Vera Rubin at the plate-measuring machine of the Carnegie Institution's Department of Terrestrial Magnetism in Washington, DC.

much faster than scientists had expected. The discovery is generally hailed as the first convincing evidence for the existence of dark matter. "It was not until Rubin's work that dark matter was confirmed," the Carnegie Institution wrote in a press release about her passing away on December 25, 2016.

The DTM, in Washington, DC, is where Ford has spent his whole career, ever since he applied for a summer job back in 1955. It's also where he helped develop the Carnegie Image Tube, an electronic device that enabled astronomers to study much fainter objects than they could with old-fashioned photographic plates. That was all decades ago.

Ellen Ford—eighty-one years old by the time of my visit— provides me with driving directions to the couple's red farmhouse in the middle of nowhere: past the Millboro Mercantile and the

Windy Cove Church, up a gravel road, and beyond a big horse barn. She welcomes me at the front porch, dressed in wellies and a windbreaker and wearing a button that says NO to the planned Atlantic Coast Pipeline. In the house, she prepares ham sandwiches with mustard—Kent's favorite.

No, it's not lonely out here, Kent Ford says, when we sit down in the living room, surrounded by the photographs from the past. But he misses the DTM lunch club, where science staff members took turns preparing meals, where hamburgers and hot dogs were allowed only once a week, and where every possible topic was up for discussion. It was during one of those lunch meetings, in 1965, that Ford and radio astronomer Bernard Burke introduced Rubin as their new colleague—the first woman on DTM's science staff, believe it or not.

It wasn't Rubin's first encounter with male dominance in science. After earning her bachelor's degree in astronomy in 1948, she wanted to go to graduate school at Princeton, but the university didn't accept female astronomy graduate students—a blatant form of gender discrimination that continued until 1975. Instead she went to Cornell and subsequently obtained her PhD in 1954 at Georgetown, where she became an assistant professor in astronomy in 1962. Even then, it was hard for her to get observing time at the big telescopes of Palomar Observatory in Southern California. There simply had never been any female observers up there before.

Upon arriving at DTM—walking distance from her home, which was convenient because the youngest of her four children was five years old in 1965—Rubin had to choose whom she wanted to share an office with: Bernie Burke or Ford. She became enchanted by the delicate parts of Ford's image tube spectrograph, which were scattered all over his desk. "She chose the spectrograph," Ford says, smiling. They shared the office for fifteen years.

This image tube spectrograph—now on display in the National Air and Space Museum on the National Mall—was the device that made Rubin and Ford's breakthrough observations possible. To study the motions of stars or nebulae, astronomers use a spectrograph, which employs either a prism or a grating to spread out light into the colors of the rainbow. Dark lines in the resulting spectrum—the "fingerprints" of various chemical elements—are slightly shifted toward the red or the blue end, depending on whether the object is receding or approaching, with the wavelength shift depending on the object's velocity. This same Doppler technique had been applied by Vesto Slipher in 1912 to detect the apparent recession velocities of galaxies, caused by the expansion of the universe (see chapter 3).

However, to record the spectrum of a faint nebula on a photographic plate, you need extremely long exposure times: sometimes up to two nights. The Carnegie Image Tube—designed by Ford and eventually manufactured by electronics company RCA—works as an image intensifier, enabling faster recording of less luminous objects. Without going into too much technical detail, a photon of light hitting the so-called cathode end of the device releases an electron. Next, a cascading process inside the vacuum tube produces ever more electrons. Eventually, the electron beam generates a glowing pixel on a phosphor screen, much brighter than the original photon. The same technology can be found in military night-vision devices.

Using the novel apparatus, exposure times of just a couple of hours were enough to record the spectra of faint objects—a huge improvement. Slipher had been the first to obtain spectra—and to derive velocities—for whole galaxies. Now, with Ford's spectrograph, it would be possible to do the same for individual objects within a galaxy, at least if the galaxy weren't too far away. This would yield valuable information about the rotational velocities within spiral galaxies, as a function of distance from the galactic center. And that, in turn, would tell you about the galaxy's mass and the way in which this mass is distributed.

A similar relation between rotational velocity and mass is found in many other flat, rotating structures in the universe, from the rings of Saturn and our solar system as a whole to protoplanetary disks surrounding newborn stars. In all these cases, just as with disk galaxies like the Milky Way and Andromeda, motions are generally governed by gravity, and velocity measurements will tell you about the distribution of mass within the rotating system.

Take our solar system, for example. If you know a planet's orbital velocity and its orbital radius (the average distance between the planet and the Sun), it's easy to calculate the Sun's mass. Thus, even if we didn't know about the Sun's size or its composition—even if we never had actually seen the Sun at all—it would be straightforward to determine its mass, just by observing the motions of the planets.

In our solar system, some 99 percent of the total mass is concentrated in the Sun itself. For a disk galaxy like Andromeda, however, the situation is a bit different: the mass is much more distributed. As a result, the orbital velocity of a star at a certain distance from the center is not determined just by the mass of the galaxy's central object (like our Milky Way, Andromeda has a central supermassive black hole) but also by the mass of all matter—visible and invisible— interior to the star's orbit. Likewise, if millions of giant planets revolved around the Sun within the orbit of Jupiter, their combined mass would increase Jupiter's orbital velocity.

Of course, you'd still expect orbital velocities to eventually fall off with increasing distance from the center. After all, the density of stars at the outer edge of a disk galaxy is much lower than the density closer to its nucleus—that's why the outer parts only show up on long-exposure photographs. So if you plot orbital velocities as a function of distance from the core, the resulting graph should show a slow decline. Such a graph is called a rotation curve. The shape of a galaxy's rotation curve yields information about its mass and mass distribution—precisely what Rubin and Ford were after in the case of the Andromeda galaxy.

The Andromeda galaxy may be the Milky Way's nearest large neighbor, but it's still 2.5 million light-years away. At that distance, it was flat-out impossible to record the spectrum of an individual star, even with Ford's powerful device. Instead the two astronomers focused on so-called HII regions (pronounced *aitch two*): luminous clouds of hot, ionized hydrogen gas, akin to the famous Orion Nebula, but much larger. These, too, orbit a galaxy's center at velocities determined by the total mass within their paths.

Starting in December 1966, the bulky image tube spectrograph was mounted on the 72-inch telescope at Lowell Observatory in Flagstaff, Arizona, for observing runs typically lasting a few nights. For each HII region, the telescope had to be precisely pointed so that the nebula's faint light would fall into the instrument, ready to be dispersed into a spectrum. A modified plate camera was used to photograph the spectrum as it appeared on the phosphor screen—back then, there was no such thing as automated electronic readout. Despite the miraculous intensification by the image tube, the 2-by-2 inch plates still had to be exposed for two to three hours. All that time, an operator had to manually steer the telescope to ensure that it would accurately follow Andromeda's slow motion across the sky as a result of the rotation of the Earth. Everything happened in a dome as frigid as the outside air, and in total darkness, lest stray light might ruin the observations.

In some cases, after the work at Lowell was completed, Rubin and Ford would load their equipment in a van for a 300-mile drive from Flagstaff to Tucson, down what later would become Interstate 17, for additional observations with the 84-inch telescope at Kitt Peak National Observatory. Finally, when all the plates had been developed, they were brought back to Washington, where Rubin meticulously measured the wavelengths of the spectral lines with a special-purpose microscope.

The black-and-white photos displayed in Ford's living room make much more sense after hearing him chat about these wonderful months in the late 1960s. A charming Rubin in a summer dress at the lower end of one of the telescopes, evidently shot during the daytime. But also Rubin wearing a thick winter coat and gloves, her eye glued to the telescope's eyepiece, apparently during one of those hour-long exposures on a cold, 7,000-foot mountain top. Rubin at the plate-measuring machine at DTM. And, of course, the resulting plot: the rotation curve of the Andromeda galaxy.

Sixty-seven HII regions, at various distances from the galaxy's core, out to some 78,000 light-years. Sixty-seven spectra, wavelength measurements, velocity determinations, and corresponding points in a diagram—the harvest of almost a year of hard work. No one had ever done something similar at this level of detail and over such a wide range of distances. And the results were a bit unexpected, for even in the outermost parts of Andromeda, where hardly any starlight was left, the rotational velocities didn't seem to fall off as expected. The rotation curve remained flat.

In December 1968, Rubin and Ford presented preliminary results at a meeting of the American Astronomical Society in Austin, Texas. Just over a year later, in February 1970, their paper "Rotation of the Andromeda Nebula from a Spectroscopic Survey of Emission Regions" was published in *The Astrophysical Journal*.[2] From their data, they concluded that Andromeda weighed in at 185 billion times the mass of the Sun, and about half of that mass was located within 30,000 light-years of the galaxy's center.

Mass and mass distribution—the original goal had been achieved. But obviously the results could only tell you about what went on within the outermost observed point. If velocities appeared to stay more or less constant out to a distance of 78,000 light-years from Andromeda's nucleus, what would happen at larger distances? How much mass could possibly hide beyond the outermost HII region in the sample?

Rubin and Ford decided not to speculate. There's no mention of dark matter at all in their 1970 paper. No reference to the earlier work of Kapteyn, Oort, and Zwicky. "Extrapolation beyond that [outermost] distance is clearly a matter of taste," they wrote.

So what about that much-circulated picture, with the rotation curve superimposed on a beautiful black-and-white photograph of Andromeda, and with data points way beyond the visible edge of the galaxy? Ford rises from the couch and slowly shuffles over to the dresser where a mounted version of the famous picture has been staring at me for over an hour. "Oh well," he says, after studying the graph in a bit more detail, "this one isn't from our first paper. It was created later. The outermost points must be radio data obtained by Mort Roberts in the mid-1970s."

It wasn't until much later that Rubin and Ford suggested that their results might hint at the existence of large quantities of "missing mass" or "non-luminous matter." In the course of the 1970s, they started to observe more distant spiral galaxies of various sizes and masses, using updated equipment on larger instruments—the almost identical 4-meter telescopes at Kitt Peak in Arizona and at Cerro Tololo in Chile.

In a paper written with Norbert Thonnard and published in the November 1978 *Astrophysical Journal Letters,* Rubin and Ford described results for ten galaxies, and concluded that the "rotation curves of high-luminosity spiral galaxies are flat, at nuclear distances as great as $r = 50$ kpc" (163,000 light-years).[3] What could that possibly mean? By that time theorists Ostriker, Peebles, and Yahil had suggested that disk galaxies are embedded in extended, massive halos (see chapter 4). Could the new results be the observational evidence for the halo model?

The authors remained cautious. "The observations presented here are . . . a necessary but not sufficient condition for massive halos," they wrote. In other words: yes, a large, more-or-less-spherical massive halo would lead to a flat rotation curve, but flat rotation curves could also be explained by additional matter just in the plane of the galaxy's disk. "The choice between spherical and disk models is not constrained by these observations."

Rubin and Ford's most famous paper appeared two and a half years later, in December 1980, again in *The Astrophysical Journal* and again with coauthor Thonnard.[4] This time, they presented observations for no fewer than twenty-one galaxies. All of them—even UGC 2885, a behemoth at least twice the size of our Milky Way—turned out to have flat rotation curves. In some cases, orbital velocities even seemed to be slightly increasing at the visible edge of the galaxy.

"The conclusion is inescapable that non-luminous matter exists beyond the optical galaxy," Rubin, Ford, and Thonnard wrote. As for the amount of this invisible stuff, they could only offer intriguing questions: "If we could observe beyond the optical image, especially for the smaller galaxies, would the velocities continue to rise . . . ? Is the luminous matter only a minor component of the total galaxy mass?"

Over the years, and with hindsight, the 1980 paper was increasingly seen as hailing a revolution in dark matter research. "Vera Rubin had transformed dark matter from a subject fit primarily for the speculators to a highly visible problem," according to astronomers Wallace and Karen Tucker in their 1988 book *The Dark Matter*.[5]

But when Rubin, Ford, and Thonnard published their results, the flat rotation curves of spiral galaxies didn't make headlines in newspapers or popular science magazines. As far as astronomy went, editors were more interested in the stunning photographs of Saturn shot by NASA's planetary explorer *Voyager 1* in November 1980.

Kent Ford retired in 1989. He and Ellen moved to their secluded red farmhouse in Millboro Springs, on the banks of the Cowpasture River. Of course, Ford stayed in touch with Rubin, meeting every now and then at parties or symposia. But in 2011, when he celebrated his eightieth birthday, she couldn't attend because of a broken hip. Rubin moved to Princeton, to live closer to her son. She was devastated when her daughter Judy died in 2014. She became forgetful; her health deteriorated. On Christmas Day 2016 came the dreaded telephone call from Carnegie's Department of Terrestrial Magnetism.

By then, dark matter—once a rather obscure astrophysical concept—had grown into one of the biggest unsolved mysteries in science, occupying the minds of hundreds of astronomers, cosmologists, and particle physicists, and Vera Rubin was seen by many as the person who had done more than anyone to bring it to the forefront of scientific research. Moreover, she had grown into a strong supporter of female scientists and an inspiration for girls interested in careers in science, technology, engineering, and mathematics.

On January 4, 2017, In a *New York Times* op-ed, Harvard physicist Lisa Randall wrote:

Of all the great advances in physics during the 20th century, surely [the presentation of convincing evidence of dark matter] should rank near the top, making it well deserving of the world's pre-eminent award in the field, the Nobel Prize. Yet to this date none has been awarded, and may never be, because

the scientist most often attributed with establishing its existence, Vera Rubin, died on Christmas Day.

Referring to other female scientists that have been overlooked for
the Nobel Prize, Randall added, "The elephant in the room is
gender."[6]

On his couch in the middle of nowhere, Kent Ford smiles his
friendly smile. "I don't have much of an opinion about that," he
muses. "I remember the DTM director saying he hoped we would
never receive the Nobel Prize, because all the publicity would leave
us no time to do any good work. Well, it didn't happen, and that's
OK." For now, Ford is happy that people are still paying attention
to the rotation curve work of more than forty years ago. "It's fun to
sit here in the countryside and to read about it in the *New York
Times.*"

Meanwhile, people shouldn't forget about the radio observations,
he says, referring to the outermost data points in the rotation curve
plot of the Andromeda galaxy that has been put back on the dresser.
"You should certainly talk to Mort Roberts."

I've already penned that name down in my notebook. Before I
leave, I also ask Ford about the 1978 PhD dissertation of Dutch radio
astronomer Albert Bosma, which is referenced in the 1980 *Astrophysical Journal* paper. "I'm sorry," he says, "I'm not familiar with
that. Vera always did most of the writing."

6

Cosmic Cartography

N SF Vera C. Rubin Observatory," reads the text on the black T-shirt of Steve Kahn, director of the Large Synoptic Survey Telescope (LSST). It's January 6, 2020, and it's the first day Kahn is wearing the shirt in public. Today, during the 235th meeting of the American Astronomical Society in Honolulu, the new name for the LSST is officially announced by Ralph Gaume, head of the National Science Foundation's Division of Astronomical Sciences. Not much later, almost every LSST employee is wearing the same shirt.

It's not the only new name Gaume announces from Room 301 of the Hawai'i Convention Center. Yes, the observatory, in northern Chile, is named after Vera Rubin, "who provided important evidence of the existence of dark matter," as NSF carefully states in an accompanying press release. But, in addition, the powerful telescope at the observatory will be known as the Simonyi Survey Telescope from now on, after an early private donor to the project. Finally, there's some consolation for astronomers who've gotten attached to the four-letter acronym LSST: henceforth, the program that will be carried out by the telescope is called the Legacy Survey of Space and Time.[1]

Vera Rubin would've been proud. With its 8.4-meter primary mirror, the Simonyi Telescope won't be the world's record holder for size, but it will be by far the "fastest" telescope in the world. Three times a week, its 3.2-gigapixel electronic eye—the largest digital camera ever built—will map the entire visible sky above the observatory. Dedicated algorithms will search through the staggering amount of data—some 20 terabytes per night—for Earth-approaching asteroids, faint supernova explosions, and lots of other transient objects in the nearby and distant universe.

Most importantly, though, the LSST survey is expected to shed new light on the mysteries of dark matter and dark energy. The program will "dramatically improve our understanding of the universe," as Kahn says. So who knows, this might well be the instrument that will finally solve the dark matter puzzle. In fact, when astronomer Anthony Tyson first came up with the idea for the grand new machine, he called it the Dark Matter Telescope.

That was back in 1996. Tyson, a researcher at AT&T Bell Laboratories in Murray Hill, New Jersey, had become the world's premier expert in weak lensing—the subtle phenomenon whereby the gravity of matter in the relatively nearby universe slightly deforms images of more distant background galaxies (we'll get back to the topic of gravitational lensing in chapter 13). If you could precisely map these tiny effects all over the sky, Tyson realized, you could deduce the distribution of gravitating matter—both visible and dark—throughout space and time. Thus, the concept for the Dark Matter Telescope was born.

It took a while before work on the telescope got started, but the project received a big boost in 2008, when Microsoft software architect, space tourist, and billionaire Charles Simonyi, through the Charles and Lisa Simonyi Fund for Arts and Sciences, donated $20 million to what would become known as the Large Synoptic Survey Telescope. Bill Gates added another $10 million. Two years later, the LSST was ranked the top priority for ground-based instruments in astronomy and astrophysics by the authoritative Decadal Survey of the National Academy of Sciences, and in 2014 the National Science Foundation secured the remainder of the funding for the futuristic telescope. The giant camera would be built by the Department of Energy's SLAC National Accelerator Laboratory Center. On April 14, 2015, at a traditional *primera piedra* ceremony, Chilean president Michelle Bachelet laid the first stone for the new facility at a mountain known as Cerro Pachón. First light is now expected in late 2023.

Cerro Pachón is located in the mountainous region east of the Chilean seaside town of La Serena. The area is home to several professional observatories, including Cerro Tololo Inter-American Observatory, Las Campanas Observatory, and the European La Silla Observatory. Thanks to the usually cloudless skies, the stable and dry atmosphere, and the low level of light pollution, this is a true astronomer's paradise. In recent decades, the region has also become a heaven for astrotourism. The signposted Ruta de las Estrellas takes you along a growing number of public observatories and stargazing sites.

The fastest way to reach the area is down Highway 41, which runs east from the Pacific Ocean into the relatively lush Valle del Elqui. However, in late June 2019, I'm driving my four-wheel-drive pickup truck on the barren mountain road D-595, slowly heading north from the small town of Samo Alto and through the Pichasca National Monument. It's a gorgeous drive through a landscape of rolling hills, dotted with small patches of vegetation and crisscrossed by deep valleys.[2]

Suddenly, between the villages of Séron and Hurtado, I'm treated to a brief but impressive view of the LSST, perched high atop a mountain ridge in the north. Next to the building I can make out a towering crane—construction of the telescope is still in full swing.

The scene can't be more than ten kilometers away as the crow flies, but to get there requires another hundred-kilometer drive, mainly on steep, sinuous gravel roads.

After crossing the Cordón Paranao mountains, I pass through the town of Vicuña, where the tourist industry is preparing for thousands of visitors who will flock here to witness the total solar eclipse on July 2. From there, it's just a fifteen-minute drive to the Control Puerta at the start of the winding and unpaved forty-kilometer access road to Cerro Pachón, which is also home to the 8-meter Gemini South Telescope and the 4.1-meter Southern Astrophysical Research Telescope.[3]

When I finally arrive at the summit, 2,700 meters above sea level, I'm blown away by the sheer size of the LSST. The cylindrical telescope enclosure ("Yes, we call it a dome," site manager Eduardo Serrano tells me) is still an open steel structure, as tall as a nine-story apartment block. But the sleek, multi-level lower part of the huge building, designed to produce as little air turbulence as possible, is completed. For now the empty control room of the telescope, at the building's top level, is a makeshift office space as well as a shelter and canteen for the construction workers. At the bottom level are the German-built coating chambers for the telescope's mirrors. Indeed, LSST's 3.4-meter convex secondary mirror would be coated with a thin reflective layer of silver just three weeks after my visit. The huge 8.4-meter primary mirror arrived on the mountain in May 2019 and will receive its aluminum coating at a later stage.

Meanwhile, the actual telescope structure has been manufactured in Spain and is ready for shipment to Chile once the dome is completed, Serrano says. "The Italian construction company which is building the dome is about two years behind schedule," he moans, looking up at the unfinished structure, silhouetted against the crystal-clear blue sky. For lack of a telescope to show off, he walks me

Artist's impression of the completed Vera C. Rubin Observatory at Cerro Pachón, Chile.

around the huge hollow concrete pier, 16 meters in diameter, which will support the 350-ton instrument. And he proudly points out the giant lift that has been constructed to transport the mirrors down from the telescope level whenever they need to be recoated.

Mapping the sky three times a week at an unprecedented level of detail will almost certainly provide lots of other results too, apart from a better sense of how mass in the universe is distributed and therefore where dark matter might be found. At least, that has always been the case of large-scale cosmic-mapping endeavors, ever since Gill and Kapteyn's *Cape Photographic Durchmusterung* and the mid-twentieth-century *Palomar Observatory Sky Survey,* which comprised almost 2,000 photographic plates of the night sky. But while earlier cosmic surveyors were mainly concerned with plotting the distribution of stars in the sky, the goal has since shifted to mapping the distribution of galaxies in the universe. Preferably in three dimensions. Or, if you include time, in four.

The effort to map the distribution of galaxies began by hand and by eye. Starting in 1948, astronomers Donald Shane and Carl Wirtanen spent eleven years meticulously counting hundreds of thousands of galaxy images on 1,390 photographic plates taken using the 20-inch Carnegie Double Astrograph at Lick Observatory on Mount Hamilton in California. Their statistical analysis of the galaxies' distribution in the sky wasn't published until 1967, and it took another ten years before Michael Seldner, collaborating with Bernie Siebers, Ed Groth, and Jim Peebles, turned the galaxy counts into a stunning image.[4] Their *One Million Galaxies* map, which has adorned the walls of astronomy departments all over the world, showed an intricate filamentary pattern, visually revealing what the statistics had indicated all along: the large-scale distribution of galaxies in the universe isn't smooth, but clumpy. How did that come about?

A two-dimensional map only gives you so much information. After all, it's just a projection of a three-dimensional reality—galaxies that appear close to each other in the sky may in fact be located at vastly different distances. To turn a 2D map into a 3D one, you not only need to know where a galaxy is located in the sky (the celestial equivalent of terrestrial longitude and latitude) but also how far away it is: you need to know its location in the third dimension.

In principle, that's easy. Remember that the light from a very remote galaxy gets redshifted by cosmic expansion, as described in chapter 3. The amount of redshift tells you how far away the galaxy is. But in practice, figuring out the distance to a remote galaxy is a tough and time-consuming job. A single photograph can yield the sky positions of thousands of galaxies at a time, but measuring redshifts requires aiming your spectrograph at each individual galaxy in turn. Moreover, to obtain a spectrum, you need a much longer exposure time than would suffice for a lone image.

In 1977, the same year in which the *One Million Galaxies* map was published, Peebles's former student Marc Davis of the Harvard-Smithsonian Center for Astrophysics (CfA) in Cambridge, Massachusetts, took up the challenge. Together with colleagues John Huchra, David Latham, and John Tonry, Davis determined the redshifts and corresponding distances of 2,400 galaxies in a rather narrow strip of sky. Huchra, an experienced observer, took almost all of the spectra, using a 1.5-meter telescope at Mount Hopkins, Arizona, and a spectrograph built with help from Carnegie's Stephen Shectman.

It took the team five years to complete this pioneering redshift survey. The resulting "slice of the universe" map, published in 1982, revealed the three-dimensional distribution of galaxies in a thin, 135-degree wedge, out to a distance of approximately 600 million light-years.[5] The map unequivocally showed that galaxies are clustered in relatively thin walls, surrounding huge, more-or-less empty voids. Studying the clustering properties of galaxies in more detail should shed more light on the formation of the large-scale structure of the universe. On whatever set this process in motion. On dark matter.

John Huchra got hooked. Teaming up with Margaret Geller, a Harvard colleague and another former Peebles student, and with

French astrophysicist Valérie de Lapparent, Huchra embarked on a much more ambitious survey of the same wedge of sky. This second CfA redshift survey, carried out between 1985 and 1995, mapped the 3D positions of no fewer than 18,000 galaxies.[6] Extremely time-consuming work, but definitely worth it. Cosmic cartography was finally coming of age.

Meanwhile, multi-object spectroscopy was coming of age, too. The idea: put an aluminum plate in the telescope's focal plane, with hundreds of small holes drilled at the precise positions where the light of the galaxies in the telescope's field of view will end up. Feed this light to the spectrograph through hundreds of glass fibers, and you can take the spectra of hundreds of galaxies in one fell swoop. Yes, you need a new "drill plate" for every new pointing, but the technique saves you an enormous amount of telescope time.

Using multi-object spectroscopy at the 3.9-meter Anglo-Australian Telescope (the same instrument Koen Kuijken and Gerry Gilmore had used in the late 1980s to prove Oort wrong), a team led by Matthew Colless of the Australian National University carried out the Two Degree Field (2dF) Galaxy Redshift Survey. Between 1997 and 2002, they determined the redshifts of a whopping 230,000 galaxies, out to a distance of some 2.5 billion light-years, using robotically positioned fibers instead of drill plates.[7] The one-million-galaxy signpost was finally coming into view: the original *One Million Galaxies* map was just two-dimensional; it would be a huge success if scientists could study the 3D positions of a similar number of galaxies.

In addition, the 2dF survey started to include the fourth dimension—time. After all, telescopes are like time machines, always providing a look into the past. Light from a distant object takes time to reach us; galaxies at distances of 2.5 billion light-years are seen as they were 2.5 billion years ago. Looking back in time enables you to study the evolution of the universe's large-scale structure. And if the growth of cosmic structure in the early days of the universe has been governed by the gravity of dark matter, deep galaxy surveys may provide telltale clues as to the true nature of the mystery stuff.

One of the most ambitious and successful 4D mapping projects to date is the Sloan Digital Sky Survey, which started in 2000 and is still running, releasing avalanches of new data almost every year.[8] The Sloan collaboration consists of hundreds of scientists from dozens of institutions all over the world. The survey employs a dedicated 2.5-meter telescope at Apache Point Observatory in New Mexico. Apart from stunning photographs, acquired between 2000 and 2009 with a huge 120-megapixel camera, the Sloan survey has yielded over four million spectra of both stars and galaxies, including so-called quasars (remote galaxies with extremely luminous cores) out to distances of billions of light-years. Quite a leap from the 2,400 galaxies of Marc Davis's first CfA redshift survey in the early 1980s.

From the Playa El Faro, at the foot of La Serena's Faro Monumental (Great Lighthouse), I look out across the Pacific Ocean. Further south, on the seaside terraces of the restaurants along the Avenida del Mar, tourists are enjoying the quiet surf and the colorful sunset. It took terrestrial seafarers centuries to explore this vast body of water and to map the many thousands of large and small islands that rise above the waves. But within the span of a mere four decades, astronomers have succeeded in mapping and studying many millions of galaxies—once described as "island universes"—in a cosmic ocean many billions of light-years across. Without ever leaving port, mind you.

Just a few years from now, the LSST will be ushering in yet another new era of cosmic cartography. During the ten-year survey, the telescope is expected to detect and image an incredible 20 billion galaxies, measuring their light output in six wavelength bands. In the case of a very remote galaxy, this energy distribution provides a rough redshift estimate—again, a proxy for the galaxy's distance from the observer—without the need to acquire a detailed spectrum. The data will enable cosmologists to reconstruct the growth of structure throughout billions of years of cosmic history. In addition the LSST will fulfill Tyson's dream of mapping the distribution of dark matter through space and time, by statistically studying the tiny deformations in the shapes of all these galaxies, caused by weak gravitational lensing.

Mapping the invisible universe. It's as if I were to climb Faro Monumental to study the undulating Pacific and use these tell-tale patterns to learn about invisible air flows, subsurface ocean currents, and hidden relief on the ocean floor. To Captain James Cook, this would seem like magic.

To be honest, I feel a little bit like that when glancing over the specs of the LSST camera and reading about the expected scientific yield of the new telescope—it's nothing less than miraculous. Thanks to its unique optical design, the 8.4-meter instrument has a field of view seven times as wide as the full Moon. The telescope is so incredibly sensitive that it won't take more than fifteen seconds to detect stars and galaxies almost a billion times fainter than what the naked eye can see. After each exposure, the massive but very stiff and compact telescope needs only five seconds to slew to the next part of the sky and start taking a new photo. Collecting some 200,000 frames per year, each containing 3.2 billion pixels, the LSST is in fact an ultra-high-definition cosmic movie camera. Apart from 20 billion galaxies, it will also detect a wealth of short-lived phenomena and moving objects like distant supernova explosions, nearby asteroids, and icy bodies in the outer solar system. Astronomers will be drinking from a fire hose for many years to come.

In which way will the Vera C. Rubin Observatory contribute to the solution of the dark matter riddle? Will the new telescope be able to fully disentangle the role of the invisible stuff in the formation and evolution of the large-scale structure of the universe? Might it even shed light on the physical properties of the mysterious substance? Only time will tell, but astronomers can't wait to find out.

On the afternoon of Tuesday, July 2, 2019, the shadow of the Moon races across the Pacific Ocean, heading east. Measuring 145 kilometers across, it sweeps over La Serena and through the Valle del Elqui at 5 kilometers per second. In the small village of Villaseca, just east of Vicuña, hundreds of tourists have gathered to witness the most impressive spectacle nature has to offer: a total solar eclipse.

Through my protective eclipse glasses, I see how the dark disk of the Moon crawls across the bright face of the Sun. Daylight seems to be slowly sucked out of the landscape. Shadows become razor-sharp; the sky turns an eerie steel blue. Dogs start to bark; birds fall silent. Below the thin remaining crescent of sunlight, Venus becomes visible. Then, all of a sudden, I am plunged into darkness, together with the telescopes at Cerro Pachón, some twenty kilometers to the southwest.

Around the ink-black silhouette of the Moon, the Sun's hot, tenuous atmosphere—normally lost in the glare of the solar disk—reveals itself as a majestic silvery-white crown of soft light. For a brief 146 seconds, the invisible becomes glaringly obvious, for all to see.

It's an overwhelmingly beautiful sight.

7

Big Bang Baryons

In March 1972 I caught the astronomy bug. I was fifteen years old. In the backyard of Wim Gielingh, an amateur astronomer who lived in the same village where I grew up, I had my first look at Saturn through a decent telescope. I never recovered. Before long, I joined a youth astronomy club. I found myself making pencil sketches of lunar craters and learning the constellations with the aid of *Norton's Sky Atlas and Reference Handbook*.

Little did I know at the time that, just then, the dark matter mystery was starting to surface. In 1972 Peebles and Ostriker were analyzing their computer simulations of rotating disk galaxies, only to discover that they couldn't be stable without lots of additional matter. Rubin and Ford started to measure the rotation curves of spiral galaxies *beyond* Andromeda, only to find that they, too, were spinning too fast. Although the astronomy magazines I was reading didn't mention dark matter yet, it slowly started to dawn on the scientific community that this was a problem that wouldn't go away.

I also didn't realize that this was the era in which cosmology was coming of age. Peebles's *Physical Cosmology* was published in the same year I had my moment of astronomical inspiration. The detection of the cosmic background radiation in 1964—a mere seven and a half years prior to my Saturn experience—had finally put some convincing observational weight behind the big bang theory.

These developments had been some fifty years in the making. Back in the first two decades of the twentieth century, American astronomers discovered that other galaxies appear to recede from our Milky Way. Soon after, Georges Lemaître in Belgium came up with the idea of an expanding universe that was born in what he liked to describe as a "primeval atom" or "cosmic egg." Science had formulated its own version of Genesis, but instead of submissively accepting this narrative as the one and only truth, astronomers soon started to flesh it out, check its implications, and subject it—if at all possible—to observational tests.

What scientists did not know at the time was that their efforts would eventually—in the early 1970s, to be more precise—lead to a wholly independent line of evidence for the existence of dark matter. Particle dark matter.

If the universe was not created 6,000 years ago by some divine being, and if it has not existed forever, you can no longer ignore the questions of why and how it has evolved into its present condition. In particular, twentieth-century physicists started to wonder about the chemical composition of the cosmos.

In 1925, in a brilliant doctoral thesis, twenty-five-year-old Cecilia Payne showed that the Sun—and, consequently, every star in the universe—is mainly composed of hydrogen, the lightest element in nature.[1] Payne may be regarded as a Vera Rubin avant la lettre. As a young woman in England, Payne studied chemistry and physics. But since Cambridge University didn't grant degrees to female students, she moved to the United States in 1923, where she became the first woman to earn a Harvard astronomy PhD.

Payne's groundbreaking work initially received a lot of opposition, notably from influential male astrophysicists. But within just a couple of years, everyone was convinced, and in the late 1930s, Carl Friedrich von Weizsäcker and Hans Bethe showed how protons (the nuclei of hydrogen atoms) could fuse into heavier helium nuclei under the enormous temperatures and pressures in the core of the Sun. In the process, energy is released, just like eminent British astrophysicist Arthur Eddington had suggested in 1920. In a very eloquent *Nature* paper, "The Internal Constitution of the Stars," Eddington wrote that "the sub-atomic energy in the stars is being freely used to maintain their great furnaces," even though he didn't yet know what the Sun was made of in the first place.[2]

Great, so hydrogen could turn into helium, the second-lightest element. Subsequent nuclear reactions could possibly produce a number of other light elements too, including carbon, nitrogen, and oxygen. But what about the many heavier chemical elements in the periodic table, like sulfur, iron, gold, and uranium? On a cosmic scale, they may not be that abundant, but still, you need to explain where they come from.

Right after the Second World War, astronomers and physicists proposed two quite different explanations. One held that nature's impressive chemical variety was the outcome of nuclear fusion in the hot and dense primordial matter. After all, conditions in the newborn universe were very similar to those in the core of the Sun, so you would expect similar reactions to take place. This view was put forward in a famous 1948 *Physical Review* publication that became known as the $\alpha\beta\gamma$ paper.[3] (The paper was released on April Fool's day, but that must have been a coincidence.)

I have heard and read about this paper ("The Origin of Chemical Elements") many times. I know the story of how Soviet-American nuclear physicist and prankster George Gamow, who wrote the paper together with his PhD student Ralph Alpher, inserted the name of his colleague and friend Hans Bethe on the author list, to arrive at something that sounded like the first three letters of the Greek alphabet. But I had never actually looked up the paper until recently. To my surprise, the $\alpha\beta\gamma$ paper is hardly more than a brief note: some 600 words, two equations, and one graph.

In contrast, the competing view, which holds that all heavy elements are produced in the interiors of stars, led to a number of very comprehensive papers, authored or coauthored by the iconoclastic British astronomer Fred Hoyle. Hoyle never believed in Lemaître's primeval atom or cosmic egg or whatever people wanted to call it. In a BBC radio interview on March 28, 1949, he even tried to make fun of the theory by calling it the "big bang," a coinage that turned out to be so apt that no one has ever come up with a better name for the beginning of the universe.

Rather than accept that all matter in the universe was created at once, billions of years ago, Hoyle believed that new hydrogen atoms were continuously being created, keeping the overall density of the universe at a constant level in spite of cosmic expansion. In this so-called steady-state model, which he published together with his Cambridge colleague Thomas Gold in 1948, there was no room for nucleosynthesis—the production of new atomic nuclei—in a primordial fireball. Instead, all atoms heavier than hydrogen would have been produced in the fiery interiors of stars.

In 1946, two years before Alpher and Gamow's $\alpha\beta\gamma$ paper, Hoyle had already penned forty pages about stellar nucleosynthesis in *Monthly Notices of the Royal Astronomical Society.* Eight years later, he provided more detail in a twenty-five-page article in *Astrophysical Journal Supplement.* Eventually he teamed up with American physicist Willy Fowler and British-American astrophysicists Margaret and Geoffrey Burbidge—a collaboration that resulted in a monumental, landmark paper in *Reviews of Modern Physics.*[4]

Published in October 1957, "Synthesis of the Elements in Stars" became generally known as the B^2FH paper, after the initials of the four authors. Another milestone in the history of astrophysics, and another quirky acronym-like nickname. But while the $\alpha\beta\gamma$ paper was just six paragraphs long, B^2FH went on for a whopping 108 pages, its thirteen chapters dense with graphs, equations, tables, and diagrams of nuclear reactions.

We now know that Alpher and Gamow were right in claiming that helium was produced during the big bang; subsequent stellar nucleosynthesis added only small amounts of helium. But the Burbidges, Fowler, and Hoyle hit the nail on the head with their assertion that elements like carbon, nitrogen, oxygen, sodium, aluminum, silicon, chlorine, calcium, and even iron are cooked up in the nuclear cauldrons in the interiors of stars—Eddington's great furnaces.

The B^2FH paper opened with an apt quote from Shakespeare's *King Lear:* "It is the stars, The stars above us, govern our conditions." They do indeed, not in the astrological way, but in the most literal sense: our very substance—the carbon atoms in our muscles, the calcium in our bones, and the iron in our blood—has a stellar origin. "We are stardust, billion-year-old carbon," as folk singer Joni Mitchell wrote in her 1969 ballad "Woodstock."

OK, so the theory of stellar nucleosynthesis, as described in the comprehensive B^2FH paper, tells us about the origin of the known, familiar world around us: the atomic building blocks of mice, motorcycles, and mountains. But in order to learn more about the universe's mysterious dark matter, we need to focus again on the brief $\alpha\beta\gamma$ publication, as Peebles realized right after the discovery of the cosmic background radiation.

This 1964 discovery by Arno Penzias and Robert Wilson (briefly described in chapter 1) was the third and final nail in the coffin of Hoyle and Gold's steady-state model. The first had been the gradual realization that the universe contains a large amount of helium—some 24 percent of the total atomic mass of the cosmos is in the form of this second-lightest element. (About 75 percent of the atomic mass of the universe is hydrogen; all other elements together comprise less than two percent.) Yes, helium is also produced in the hot interiors of stars, but not in such huge quantities—only big bang nucleosynthesis fits the bill.

The second line of support for the big bang theory came in the early 1960s. Radio astronomers discovered that galaxies in the very distant universe have different properties from galaxies in our cosmic neighborhood. Since a remote galaxy's light may take billions of years to reach us, this observation means that galaxies looked different in the past than they do now. In other words: the universe is not in a steady state but is indeed evolving, in agreement with the big bang theory.

The discovery of the cosmic background radiation—also known as the cosmic microwave background, because of its peak wavelength of approximately one millimeter—finally closed the case in favor of the big bang: the radiation is rightfully known as the "afterglow of creation." (I'll get back to the peculiarities of the cosmic microwave background in chapter 17.) When Bob Dicke asked his Canadian postdoc, "So Jim, why don't you delve into the theory behind all this?" Peebles knew that sinking his teeth into the early universe would shed light on the chemical makeup of the cosmos. Apart from working on the propagation of over- and underdensities in the hot, viscous soup of particles and radiation, he also studied the way in which the outcome of nuclear reactions in the first few minutes after the birth of the universe, and in particular the amount of deuterium, depends on the ever-decreasing cosmic matter density.

This is not a book about the big bang, so don't expect me to go into too much detail, but what matters most here is that elementary particles known as quarks combined into the first nucleons (protons and neutrons, the constituents of all atomic nuclei) when the universe was about one second old. Because of their mass, protons and neutrons are also known as baryons, from the Greek word meaning "heavy."

At first, the number of protons was almost equal to the number of neutrons, but before long, that started to change dramatically. Because of the extreme temperatures of the newborn universe, solitary baryons could not yet combine into nuclei. And while the neutrons in an atomic nucleus are stable, free neutrons slowly but surely decay into protons. Within a time span of just a couple of minutes, the number of neutrons decreased significantly, while protons became ever more abundant.

By the time temperatures had dropped to a billion degrees or so (low enough for atomic nuclei to start forming), only one-eighth of the total baryonic mass in the universe was in the form of neutrons; the rest was protons. Through various nuclear reactions, most of the neutrons eventually ended up in helium nuclei, which consist of two protons and two neutrons. The remaining protons were left behind as nuclei of hydrogen atoms. Almost no nuclei heavier than helium could form, and after a brief but extremely energetic period of nucleosynthesis, temperatures and densities in the early universe dropped even further, and fusion reactions came to a halt.

It's now quite straightforward to calculate the outcome of this primordial nuclear mayhem. If you do the math properly, you'll find that about three quarters of the total baryonic mass of the universe is in the form of hydrogen, while something like one quarter of the mass is in the form of helium, in satisfying agreement with measurements of the cosmic abundances of these two elements. In other words: big bang nucleosynthesis provides a neat explanation for the observed chemical composition of the universe—something that the steady-state model (or divine creation, for that matter) fails to accomplish.

So what about deuterium and cosmic density? Where does Peebles's work come in? Well, helium nuclei don't form in one fell swoop. It's not like two solitary protons and two solitary neutrons happen to bounce into each other at exactly the same time. Instead, there are a number of possible intermediate steps, deuterium being the most important and the one we're focusing on here.

While a hydrogen nucleus is just one proton, a nucleus of deuterium (sometimes called a deuteron) consists of one proton and one neutron, bound together by the strong nuclear force. It's still hydrogen—chemical elements are defined by the number of protons in their nuclei—but it's almost twice as massive, hence the common name "heavy hydrogen." Pretty soon after their formation, deuterium nuclei take part in yet another range of nuclear reactions, eventually leading to the production of helium. But there's not much time for these reactions to occur: as a result of the expansion of the universe, temperatures are dropping rapidly, and big bang nucleosynthesis stalls, with the result that some deuterium is left unused.

In a November 1966 paper in *The Astrophysical Journal,* Peebles showed that the relative amount of deuterium in the universe critically depends on the density of nuclear matter (baryonic matter) during the brief epoch of nucleosynthesis.[5] The higher the density, the more efficiently the nuclear reactions proceed, and the less deuterium is left over as a residue. In contrast, a lower density during this critical epoch results in a higher deuterium abundance. Similar arguments hold for some other rare atomic nuclei, including helium-3, which contains two protons but just one neutron, but the dependency is most pronounced for deuterium.

Measuring the current abundance of deuterium in the universe would thus provide information about the cosmic density during big bang nucleosynthesis. And extrapolating from that, it's not hard to calculate the current density of baryonic matter in the universe, after billions of years of cosmic expansion. In other words: precise deuterium measurements can tell you the average density of "normal" matter, mainly consisting of atomic nuclei (baryons).

Five months after Peebles published his calculations, they were confirmed in yet another (and much more detailed) *Astrophysical Journal* paper by Robert Wagoner and Willy Fowler of Caltech, joined by Hoyle, who was still critical of the big bang but nevertheless made important theoretical contributions to the idea.[6] In January 1973 Wagoner, then at Cornell, published an updated account, "Big-Bang Nucleosynthesis Revisited."[7] It appeared that scientists had really solved the thorny issues of the origin of light elements in the primeval fireball. Now they only needed to gauge the relative deuterium abundance, and they would know the current baryonic mass density of the universe.

Directly measuring the cosmic abundance of deuterium is hard, but space science came to the rescue. On August 21, 1972, just a few months after I had my first telescopic look at Saturn, NASA launched its third Orbiting Astronomical Observatory (OAO-3), nicknamed Copernicus after the Polish astronomer whose 500th birthday was approaching in early 1973. One of the Copernicus observatory's instruments was an 80-centimeter ultraviolet telescope-plus-spectrometer, developed at Princeton. Detailed ultraviolet spectra of bright stars (which cannot be acquired from the ground) would reveal absorption lines of both interstellar hydrogen and deuterium: these filter out certain wavelengths of ultraviolet light. From the relative amounts of absorption, the abundance ratio of deuterium could be calculated.

In December 1973, Peebles's Princeton colleagues John Rogerson and Donald York published the first results in *The Astrophysical Journal,* based on Copernicus observations of the bright southern-hemisphere star Agena (Beta Centauri).[8] Their conclusion: inter-

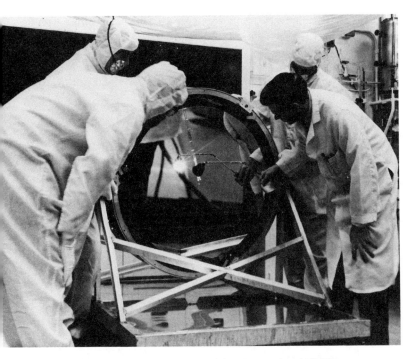

Technicians testing the 81-centimeter mirror of the telescope aboard NASA's Copernicus satellite. When launched in 1972, it was the largest space telescope ever flown.

stellar space contains just one deuterium nucleus for every 70,000 hydrogen nuclei. A tiny portion left over from the nuclear reactions in the newborn universe. From this value, the amount of baryonic matter in the universe—atomic nuclei, basically—could be derived.

Using the latest calculations by Wagoner, Rogerson and York arrived at an average cosmic baryon density of 1.5×10^{-31} grams per cubic centimeter. A 1976 paper based on ultraviolet observations of four more stars (including Spica, the brightest star in the zodiacal constellation Virgo) yielded more or less the same result.[9] Finally astronomers had a reliable estimate for the total amount of "normal" matter in the universe. And the result had dire implications for our ideas about dark matter.

Peebles, Ostriker, and Yahil were well aware of the Copernicus results when they wrote their 1974 paper "The Size and Mass of Galaxies, and the Mass of the Universe." As you may recall from chapter 4, they used a large variety of dynamical observations and arguments to determine the average mass-to-light ratio of galaxies. Then, by estimating the number of visible galaxies in a certain volume of space, they calculated the average mass density of the universe, arriving at a value of 2×10^{-30} grams per cubic centimeter—some thirteen times denser than what Rogerson and York found.

If the current density of the universe was really so much higher than the value inferred by Rogerson and York, then the density also must have been higher during the epoch of big bang nucleosynthesis. That would have resulted in a much smaller cosmic abundance of deuterium than the Copernicus measurements had indicated.

Unless.

There you go again. Unless most of the mass in the universe does not consist of baryons. The deuterium abundance only tells us about the density of atomic nuclei, the building blocks of "normal" matter that take part in nuclear reactions. What if the universe contains lots of abnormal matter? Matter that does not consist of atomic nuclei? Nonbaryonic matter? That could explain the dynamical observations without conflicting with the deuterium measurements.

It seems like a large leap of faith, and Ostriker, Peebles, and Yahil certainly didn't jump to conclusions. Maybe some other process could have produced additional amounts of deuterium in the billions of years since the big bang, although no one has ever come up with a viable mechanism. Or maybe their estimates for the number of galaxies in the local universe were too high. Indeed, an independent analysis by Richard Gott, James Gunn, David Schramm, and Beatrice Tinsley, also published in 1974, arrived at a somewhat less disturbing result.[10] But over the years, it slowly started to dawn upon cosmologists that nature really was telling us something important. If, for whatever reason, the total mass density of the universe is significantly higher than the value derived from big bang nucleosynthesis calculations, then we have to accept that there's a lot of nonbaryonic matter out there. Dark and weird.

Kapteyn, Oort, and Zwicky; Peebles and Ostriker; and Rubin and Ford—they all found circumstantial evidence for the existence of substantial amounts of dark matter. But their results did not, in and of themselves, imply that "dark" means "mysterious." Their findings could be explained by large amounts of dim dwarf stars, huge clouds of cold interstellar gas, or even a population of invisible black holes. That changed as the theory of the big bang became more convincing. Across billions of years of cosmic history, the big bang itself brought forward a new courtroom exhibit in the dark matter trial: not some kind of astrophysical object, but a strange, unknown particle. Suddenly, the old ideas—the easy explanations— appeared to be off the table. Nonbaryonic dark matter—we'd better get used to it.

Gott, Gunn, Schramm, and Tinsley started their comprehensive paper on the mass density of the universe with a quote from the Roman poet and philosopher Lucretius: "Desist from thrusting out reasoning from your mind because of its disconcerting novelty. Weigh it, rather, with a discerning judgment. Then, if it seems to you true, give in."

Right—what else could one do but give in?

They might as well have added an unsettling *Star Trek* quote: "Resistance is futile."

8

Radio Recollections

Frustrated?

Albert Bosma only has to think for a split second. Then he decisively replies, "I'm not easy to frustrate. But I keep notes of everything. Who knows, I might write a book someday."

Most of his radio astronomer colleagues, as well as dark matter historians, would agree that Bosma's discoveries in the mid-1970s were the first to really clinch the case for the existence of dark matter in galaxies. But beyond professional circles, he is hardly known. Instead, Vera Rubin's name is all over the place. How could Bosma not be frustrated?

I am meeting the short, long-haired, and bearded astronomer in the support building of the Westerbork radio observatory, in the sparsely populated Dutch province of Drenthe.[1] The observatory is built next to a former Nazi transit camp. Between 1942 and 1944, a staggering 97,776 Jews, Roma, and Sinti—including young Anne Frank and her family—were deported by train from here to Auschwitz and Sobibór, where almost all of them were killed. Today the area is used to peacefully study the mysteries of the universe. It's a cold and drizzly November morning, but every now and then the Sun breaks through the clouds, spotlighting the kilometers-long row of fourteen 25-meter radio dishes outside.

Well past retirement age, Bosma is still an active radio astronomer at the Laboratoire d'Astrophysique in Marseille, France. He just returned from a trip to China and is now visiting his family in the Netherlands. Drenthe is where young Albert grew up, in the small village of Smilde. It's where his math teacher Dr. Knol sparked his interest in astronomy. And the Westerbork observatory is where his groundbreaking observations were carried out, as part of his PhD research at the University of Groningen. The observations many people have never heard about.

Yes, you could choose to fight the battle, he says, but you would end up doing nothing else.

One thing's for sure: the 1970 Rubin and Ford paper on the Andromeda galaxy, described in chapter 5, didn't prove the existence of dark matter. It couldn't. As Australian astronomer Ken Freeman and science teacher Geoff McNamara wrote in their 2006 book *In Search of Dark Matter,* "Flat optical rotation curves can rarely provide conclusive evidence for dark matter, because they do not reach out far enough from the center of the galaxy."[2]

Neither were Rubin and Ford the first astronomers to note that something peculiar was going on with the rotation of disk galaxies. More than thirty years earlier, Horace Babcock—who would become director of Palomar Observatory in 1964—already found that the edge of the visible disk of Andromeda rotates faster than astronomers expected.[3] Walter Baade and Nicholas Mayall obtained similar results in 1951.[4]

Moreover, in the same year Rubin and Ford published their Andromeda results, Freeman, too, discovered that some galaxies appeared to contain more matter than you would guess on the basis of visual observations. Freeman analyzed the distribution of starlight in the disks of thirty-six galaxies and derived the rotation curves expected of those systems, assuming they contained only stars. Back then, rotational data were available for just a handful of galaxies, and in at least two cases, M33 and NGC 300, the measured rotation curves deviated from the computed ones.[5] "If [the data] are correct," Freeman wrote in his *Astrophysical Journal* paper, "then there must be in these galaxies additional matter which is undetected. . . . Its mass must be at least as large as the mass of the detected galaxy, and its distribution must be quite different from the . . . distribution which holds for the optical galaxy."

So yes, something surprising was going on, but nothing weird enough to convincingly show that galaxies are embedded in large dark matter halos. After all, optical observations could only reveal the mass distribution in the visible part of a galaxy. As Rubin and Ford had noted in their 1970 paper, "Extrapolation beyond that distance is clearly a matter of taste."

Or a matter of wavelength. For beyond the apparent edge of a galaxy lie tenuous, invisible clouds of cold hydrogen gas that can only be observed at radio wavelengths.

From Kent and Ellen Ford's farmhouse in Millboro Springs, it's another sixty miles farther northwest, through the Allegheny Mountains and into Pocahontas County, West Virginia, to the venerable Green Bank Observatory. Originally run by the National Radio Astronomy Observatory (NRAO), Green Bank harbors the largest fully steerable radio dish in the world. It is also a Valhalla for science historians.[6]

Just beyond the entrance of the observatory grounds is a full-scale replica of physicist Karl Jansky's 30-meter-diameter "merry-go-round" antenna, which made the first detection of cosmic radio waves back in 1931. Across the entrance road is the refurbished and freshly painted 9.4-meter dish constructed six years later by radio engineer Grote Reber in his mom's backyard—the first instrument ever to yield a crude map of the radio sky. And in the observatory's

Residence Hall Lounge, a plaque commemorates the fact that here, in November 1961, staff astronomer Frank Drake presented his famous eponymous equation. (The Drake Equation estimates the number of extraterrestrial civilizations in our Milky Way galaxy from whom we could potentially detect radio emissions.)

One slightly less conspicuous exhibit on site—in fact the one that enabled the later observations by Bosma and his radio astronomy colleagues—is the original horn antenna used by Harvard grad student Doc Ewen and his thesis advisor Edward Purcell to detect so-called line emission of neutral hydrogen—radio waves with the very specific frequency of 1420.4 megahertz, corresponding to a wavelength of 21.1 centimeters. This hallmark discovery, made on March 25, 1951 (Easter Sunday; Ewen was working 24/7 at the time), eventually made it possible to map the outer parts of distant galaxies. And since the Doppler effect works the same way for radio waves as it does for visible light, precise observations of the 21-cm line reveal the kinematics of the hydrogen gas, including the rotation of gas clouds way beyond the visible disk of a galaxy, where stars are all but absent.

Ewen and Purcell's search had been motivated by a prediction from Dutch astronomer Henk van de Hulst, a student of Oort's at Leiden. In 1944, during the Second World War, van de Hulst—the son of a famous Dutch author of children's books—was asked by his visionary thesis advisor to check if the newly discovered cosmic radio hiss might somehow contain information on the omnipresent cool neutral hydrogen in interstellar space. After some literature study and handwritten arithmetic, the twenty-five-year-old student concluded that there should be a faint hydrogen signal at a wavelength of 21 centimeters.

Right after the war, the Leiden group turned a 7.5-meter radar antenna (left behind by the Germans) into a radio telescope and started searching for the 21-cm line. But partly due to delays resulting from a fire in their receiver, they only succeeded some seven weeks after the discovery by Ewen and Purcell, who knew about van de Hulst's prediction. Not much later, Australian radio engineers Chris Christiansen and Jim Hindman obtained a third independent detection; all three results were published in the same September 1, 1951, issue of *Nature*.[7]

By then, Oort was busy gathering funds for what would briefly be the world's largest radio dish: the 25-meter Dwingeloo telescope. Inaugurated in April 1956, the instrument would make history by providing the first detailed map of the spiral structure of our own galaxy. However, the first 21-cm observations at Dwingeloo were not targeting our Milky Way but rather Andromeda. Under the supervision of van de Hulst, astronomers Hugo van Woerden and Ernst Raimond obtained the first ever rotation curve of another spiral galaxy based on HI observations. (HI is neutral hydrogen; as you may recall from chapter 5, HII is ionized hydrogen.)

Those were the early days of radio astronomy. For each individual fifteen-minute measurement, the huge dish and the bulky line receiver were prepared manually. Pointing corrections had to be calculated by hand. Data were written on rolls of paper by a pen recorder. When they weren't observing or tweaking the hardware, van Woerden and Raimond slept in the guest room of Lex Muller, the telescope's designer and administrator, whose house was right next to the dish. Mrs. Muller provided breakfast, lunch, and dinner.

The Dwingeloo results on Andromeda were published in November 1957, in the *Bulletin of the Astronomical Institutes of the Netherlands*.[8] In a 2020 interview shortly before his death, van Woerden recalled that, indeed, rotational velocities hardly appeared to be declining with increasing distance from the galaxy's center, but back then, no one was too surprised by that. "It wasn't an issue at all," he said. That started to change only slowly, when other astronomers obtained more detailed radio observations of our nearest galactic neighbor, as well as HI rotation curves for other disk galaxies.[9]

One of those astronomers was Seth Shostak, who would later become senior astronomer at the SETI Institute in Mountain View, California.[10] (SETI stands for the Search for ExtraTerrestrial Intelligence. No, nothing has been found yet.) In the late 1960s, as part of his PhD research, Shostak spent a lot of time at Caltech's Owens Valley Radio Observatory near Big Pine, California, close to the Nevada border. There he studied the distribution and dynamics of neutral hydrogen in three galaxies, including NGC 2403, which is some 3.5 times farther away than Andromeda.[11]

By the time Shostak was doing his research, radio astronomers had constructed telescopes much larger than the 25-meter Dwingeloo dish in the Netherlands or those at Owens Valley. Jodrell Bank Observatory in northern England operated a 76-meter dish (now known as the Lovell telescope, after the observatory's first director, Bernard Lovell); Australia had a 64-meter instrument at Parkes, New South Wales, nicknamed simply The Dish; and Green Bank held the world record with its giant 90-meter telescope. But what the smaller antennas at Owens Valley lacked in size and sensitivity, they made up for with their flexibility and larger angular resolution, a measure of the amount of detail that can be seen.

The two identical antennas Shostak used were each 27.4 meters in diameter, but they could be moved on rails, and the data they collected were precisely combined ("correlated") into one set of observations, as if the two instruments were small parts of one huge virtual dish. This kind of system, called an interferometer, not only has a higher angular resolution than a single-dish instrument but is also much more efficient. Radio telescopes generally have a very small field of view, as if you're watching the sky through a drinking straw, so you can create a larger "picture" only by carrying out many observations in succession, usually over a period of many days or even weeks. In contrast, an interferometer can build up a two-dimensional radio image in less than a day, through a process known as aperture synthesis.

More often than not, Shostak was the only person at the observatory, spending long nights in the control room listening to the eerie grinding sounds of the antennas outside. He couldn't help thinking about the multitude of galaxies, stars, and planets in the wider universe and about the possibility that some of those distant worlds might be populated by alien civilizations. Wouldn't it be great to use radio telescopes to eavesdrop on their interstellar communications? In a 1959 *Nature* paper, physicists Giuseppe Cocconi and Philip Morrison had suggested that very idea.[12] If you know where to look, it's easy to see when Shostak first became passionate about SETI. At the end of his 1972 dissertation, he wrote, "This thesis is dedicated to NGC 2403 and its inhabitants, to whom copies can be furnished at cost."

Between 1971 and 1973, Shostak and his thesis advisor David Rogstad, who had moved to Groningen in the Netherlands to work with the new Westerbork telescope, published papers on a total of six galaxies, including NGC 2403, M101 (also known as the Pinwheel galaxy), and M33—the third major member of the so-called Local Group, to which our Milky Way and Andromeda belong. In each case, they found that clouds of cold hydrogen gas way beyond the optical edge of the galaxy were rotating much faster than expected, indicating the presence of "low-luminosity material in the outer regions of these galaxies," as they wrote in a September 1972 paper in *The Astrophysical Journal*.[13]

Meanwhile, NRAO astronomer Morton Roberts was studying neutral hydrogen in the Andromeda galaxy, using the then-largest radio telescope in the world—the Green Bank Telescope, which had become operational in 1962. Improving on the pioneering Dwingeloo observations by van de Hulst, Raimond, and van Woerden, Roberts published his first results in 1966—just one year after Vera Rubin started to share an office with Kent Ford at Carnegie's Department of Terrestrial Magnetism.[14] Roberts' paper is referenced in Rubin and Ford's 1970 publication on Andromeda. "I knew Vera pretty well," Roberts tells me during a Zoom interview from his home in Alexandria, Virginia. "She was a very kind person, and happy to have found a male astronomer listening to her."[15]

In the early 1970s—when he started to get better and better results for Andromeda, culminating in a 1975 paper with Robert Whitehurst—Roberts gave Rubin a telephone call.[16] "I have something interesting for you," he told her. "Are you around this week?" A few days later, he drove the 120 miles from NRAO's headquarters in Charlottesville, Virginia, to Carnegie's DTM lab in Washington, DC, where he met with Rubin, Ford, their colleague Norbert Thonnard, and Sandra Faber, a Harvard PhD student who lived in DC and had been offered a temporary DTM desk by Rubin.

"Please get me a copy of *The Hubble Atlas of Galaxies*," Roberts asked Faber. She duly went off to the library to fetch the iconic 1961 book, famous for its beautiful black-and-white photographs of dozens of galaxies. Back in the meeting room, Roberts opened the atlas to the page featuring the Andromeda spiral. Using tracing paper, he plotted his newest hydrogen velocity measurements, out to a distance of some 95,000 light-years—far beyond what was pictured in the book. Even there, the galaxy's rotation curve remained flat. Everyone in the room fell silent. When Faber asked, "So what? What's the significance of a flat rotation curve?" they all swiveled toward her. "Don't you see? There's no light there!"

From then on, plots of the rotation curve of the Andromeda galaxy usually contained Roberts's HI velocities in the outer parts of the galaxy, although he believes the famous image with the photo of Andromeda in the background—the one I saw at Kent Ford's house—wasn't published until 1987.

Whether or not Rubin, Ford, and Thonnard were also aware of the work by Rogstad and Shostak remains unclear—the trio did not reference it in their 1978 and 1980 publications. But Shostak just can't imagine they didn't know about it. "In 1972, thanks to Mort Roberts, I had a postdoc job at NRAO," he says. "One of the summer students there was Vera's twenty-year-old daughter Judy, who would later become an astronomer herself. I'm sure she must have discussed our work with her mother. Vera didn't have flat rotation curves until a couple of years later; we had done it years before."

But radio astronomy was a novel and unfamiliar technique to most astronomers, and not too many people took the results all that seriously back then, according to Roberts. "Some were outright skeptical," he says. "Most were at least very cautious. Certainly no one made the connection to the earlier work on galaxy clusters by Fritz Zwicky." Dark matter still had a lot of converts to win.

As it turned out, the big game changer—at least as far as rotation curves go—was the Dutch Westerbork Synthesis Radio Telescope, an interferometer like the one in Owens Valley but with fourteen dishes instead of two, each 25 meters in diameter.[17] Inaugurated on June 24, 1970, initially with twelve antennas, this new brainchild of Jan Oort attracted a lot of scientists from abroad, most of whom ended up having their office at the Kapteyn Astronomical Laboratory at the University of Groningen. Rogstad spent a couple of years there instructing students—including Albert Bosma—in data analysis and computer technology. Roberts went to Groningen to work on 21-cm observations of the spiral galaxy M81 with Arnold Rots. And Shostak arrived in 1975. He stayed for thirteen years.

Bosma once described the Groningen radio astronomers as "the young Turks of the Kapteyn lab." At a 1973 conference in Cam-

The Westerbork Synthesis Radio Telescope in the Netherlands, used by Albert Bosma to measure rotational velocities in the outer regions of spiral galaxies.

bridge, England, their younger colleagues shamelessly boasted that the Westerbork telescope was much better than the older Cambridge Interferometer. Eminent British radio astronomer and observatory director Martin Ryle, who more or less had invented aperture synthesis, wrote a formal letter to Oort, complaining about the group's tacky behavior. Subsequently, the "Westerbork cowboys" were allowed to attend a conference in Onsala, Sweden, only if chaperoned by an older staff member, lest things get out of hand once more.

Then again, the Westerbork telescope was indeed very powerful, as Bosma's PhD research showed. Bosma took advantage of a new piece of technology: inspired by Rogstad, Groningen astronomer Ron Allen had developed a novel spectrometer that enabled eighty simultaneous observations at slightly different wavelengths. Using this 80-channel filter bank receiver, Bosma studied one galaxy after another, measuring Doppler shifts both within and well beyond the optical edge, charting velocities all over the place, mapping the extended distribution of neutral hydrogen, and discovering warps in the outermost parts of the hydrogen disk.

It all seemed to confirm the earlier, less precise, and less sensitive results of Rogstad and Shostak and of Roberts and Whitehurst. Eventually, no fewer than twenty-five galaxies turned out to have flat rotation curves out to very large distances from their cores, indicating the presence of large amounts of invisible mass way beyond the optical disk. Bosma presented initial results at conferences in 1976 and 1977, but the full extent of his work only became clear with the publication of his 1978 dissertation "The Distribution and Kinematics of Neutral Hydrogen in Spiral Galaxies of Various Morphological Types."[18] Later that year, Rubin, Ford, and Thonnard described their results for just ten galaxies, based on optical observations.

So, is Albert Bosma frustrated?

Well, at least it's remarkable to read so many different stories, he says. For instance, in her 2014 book *The Cosmic Cocktail,* theoretical astrophysicist Katherine Freese writes, "It was the work of Rubin and Ford that clinched the case for dark matter in galaxies. Their observations persuaded astronomers that dark matter must exist . . . and the two deserve a Nobel Prize for this discovery."[19] Likewise, in her *Nature* obituary of Rubin, Princeton astrophysicist Neta Bahcall calls her "the 'mother' of flat rotation curves and dark matter" and states that "her groundbreaking work confirmed the existence of dark matter and demonstrated that galaxies are embedded in darkmatter halos, which we now know contain most of the mass in the universe."[20]

Yes, Bosma wrote a letter to the editor in response to Bahcall's obituary.[21] The letter explains why Bahcall's piece "oversimplifies the dark matter problem": you really need radio data to probe the outermost regions of galaxies. However, it's impossible to respond to every incomplete or biased publication—there's so much interesting radio astronomy to do. "There's a lot of reinterpretation going on," he says, "and much of it is outright wrong. I just keep track of what everybody's thinking. But I'm a bit hesitant about writing that book—I just haven't got enough time."

Shostak has mixed feelings, too. "I have a big soft spot for Albert," he tells me, but "I'm not terribly upset." Still, "It's true that Vera came late to the party. All that talk about a Nobel Prize, and now a large telescope being named after her . . . it makes you feel kind of strange." Then again, he adds, she never claimed priority herself. Indeed, as I have noted before, Bosma's thesis is referenced in the 1980 publication by Rubin, Ford, and Thonnard. And in their 1978 paper, the authors make clear that "Mort Roberts and his collaborators deserve credit for first calling attention to flat rotation curves."

Sandra Faber, who later became a distinguished professor at the University of California, Santa Cruz, believes that—contrary to what is usually the case—Rubin's current record in history has actually been helped by the fact that she was a woman. It is a remarkable example of reverse gender inequality. "Bosma's thesis is brilliant. Two hundred years from now," she muses, "people will certainly realize how important his contributions have been."

Hopefully, it won't take that long.

PART II

Tusk

9

Into the Cold

By now you may be wondering when this book will leave the past for what it is and turn to the present. After all, we're at one-third of the story, and we still seem to be stuck in the 1970s. But don't worry—we'll get there. If you want to understand the quest for solutions to the dark matter riddle, you first need to know how the problem arose in the first place. And although the mystery is almost a century old, as we saw in chapter 3, it just so happens that almost all of the most important developments took place in the roaring seventies.

As a quick recap, we've learned that galaxies can't be stable, unless they're embedded in giant, massive halos. Moreover, galaxies are much more massive than you would guess on the basis of their visible content. Rotational velocities do not decrease with increasing distance from the galaxy's center but remain more or less constant—a sign that there is more matter in galaxies than is apparent through telescopes. The relative smoothness of the cosmic microwave background suggests that, in the moments after the big bang, weird particles must have already started to form a dark, massive scaffolding that would only later pull in the familiar baryonic matter. Finally, the big bang cannot have produced enough baryonic matter to explain the dynamical observations and the growth of cosmic structure, indicating that most of the gravitating mass in the universe must be in some unfamiliar, nonbaryonic form.

If this all seems confusing to you, even with the benefit of hindsight, can you imagine how confusing it must have been for scientists in the 1970s? Not everyone was equally aware of every single piece of evidence; some observations were more reliable than others; and astronomers who were used to studying stars and galaxies with optical telescopes suddenly had to wrap their head around the intricacies of radio astronomy and particle physics. Little wonder then that ignorance, caution, and skepticism were all over the place. A lot of things were going on at the same time, and many scientists chose to sit on the fence, waiting for the dust to settle.

That began in 1979, thanks to an influential fifty-two-page review paper in *Annual Reviews of Astronomy and Astrophysics,* written by Sandra Faber and John Gallagher.[1] Titled "Masses and Mass-to-Light Ratios of Galaxies," the paper provided a neat wrap-up of all the available evidence for the existence of dark matter, including the observations by Albert Bosma described in the last chapter. The authors focused on galaxy dynamics and masses—the very first sentence reads, "Is there more to a galaxy than meets the eye (or can be seen on a photograph)?" Faber and Gallagher wrote that they were "especially concerned with the current status of the 'missing mass' problem" and reached an outspoken conclusion: "After reviewing all the evidence, it is our opinion that the case for invisible mass in the Universe is very strong and getting stronger."

Some eight years earlier, as you may recall, Faber was the graduate student who was unimpressed by flat rotation curves, when Morton Roberts was showing his Andromeda galaxy radio results to Vera Rubin. After moving to the University of California at Santa Cruz and to its Lick Observatory on Mount Hamilton in 1972, Faber continued her Harvard PhD research on the dynamics and evolution of elliptical galaxies—huge oblate or prolate collections of randomly moving stars.[2]

Elliptical galaxies lack a distinct, orderly rotating disk. Neither do they contain outlying clouds of neutral hydrogen. As a result, it's much harder to plot rotation curves for them. Instead, Faber studied their velocity dispersion—a measure of the spread of stellar velocities within an elliptical. And although the case for invisible mass around elliptical galaxies was less convincing than for spirals, there were several lines of evidence that supported the view that ellipticals, too, were embedded in huge dark matter halos.

Faber's interest in the dark matter mystery only got stronger when Ostriker visited Santa Cruz to talk about his work with Peebles and Yahil. Eventually she teamed up with University of Illinois astronomer John Gallagher—a former student of Ostriker's who later would become the director of Lowell Observatory in Flagstaff, Arizona—to write the 1979 paper for *Annual Reviews.* This was the publication that everyone had been waiting for: clear, factual, very readable, authoritative, complete, and with firm conclusions. The astronomy community was finally convinced that dark matter was real. As the authors stated, "No valid alternative explanation has been put forward." Before long, dark matter made its appearance in college textbooks.

Five years later, in October 1984, Faber was coauthor on yet another seminal paper, in *Nature.* This time, the main question was not "Is dark matter real?" but "What kind of dark matter are we talking about?" Moreover, the paper discussed the role of dark matter in the origin of the large-scale structure of the universe and in the formation of galaxies. In less than a decade, the mystery stuff had evolved from some weird, hypothetical cosmic ingredient to the master architect of the physical world, without which there may not have been galaxies, stars, planets, or people like you and me.

The question "What kind of dark matter?" may surprise you at first sight. Hadn't we just concluded that it has to be nonbaryonic? After all, the big bang never produced enough atomic nuclei to explain the current mass density of the universe as derived from dynamical arguments, right? True enough, but nonbaryonic particles can have a wide variety of properties. For one, they can be very massive (as particles go), which would mean they would move relatively sluggishly, or they can be extremely lightweight, in which case they would zip around at close to the speed of light.

For instance, electrons are low-mass, fast-moving particles, and since they are not composed of quarks, they don't officially belong to the baryon family. Obviously, dark matter can't consist of electrons, which are negatively charged particles: if dark matter particles had an electrical charge, we would have discovered them long ago. But we also know of a nonbaryonic particle that does not have an electrical charge: the neutrino. Could neutrinos comprise the dark matter?

You will read more about neutrinos in chapter 23. For now, all you need to know is that they are not part of atoms and molecules and that they must have been produced in copious amounts during the big bang. Whether neutrinos could comprise dark matter depends on whether they have mass, even a teeny-weeny bit. If so, their sheer numbers could explain the high mass density of the universe, as inferred from galaxy dynamics.

Before delving into the mass of neutrinos, I need to provide a little bit of background on particle masses in general. Physicists usually express particle masses in units of energy. After all, according to Albert Einstein's famous equation $E = mc^2$, mass (m) and energy (E) can be converted into one another. For instance, the mass of an electron (9.11×10^{-31} kilograms) equals an energy of 511,000 electron-volt (eV). In contrast, protons are much more massive: one proton weighs in at 1,836 times the mass of an electron, corresponding to 938.3 million eV (MeV). (One electronvolt, by the way, is the amount of kinetic energy gained by a single electron if it is accelerated through an electric potential difference of one volt; it equals 1.6×10^{-19} joule.) If this is too much detail for you, just remember that a proton weighs very little (about one-trillionth of the mass of a typical bacterium), and an electron is almost 2,000 times less massive still. But what about neutrinos?

Well, neutrinos freely stream through the universe without us ever noticing, and according to the Standard Model of particle physics, their mass should in fact be zero. Then again, scientific theories have been wrong before.

Around 1980, the main champion of massive neutrinos as dark matter candidates (massive in the sense that they have some non-zero mass; not in the sense that they are particularly heavy!) was theoretical physicist Yakov Zeldovich. Zeldovich had been a key player in the Soviet nuclear weapons program during the Second World War. He was also one of the first to calculate how the current large-scale structure of the universe—clusters, superclusters, and voids—may have resulted from small density fluctuations in the primordial soup, solely through the power of gravity. Might this process have started with neutrinos gathering together, he wondered? Could neutrinos be the dark matter?

Just five kilometers southeast of Zeldovich's Moscow State University, at the Institute for Theoretical and Experimental Physics, his colleagues Valentin Lyubimov and Evgeny Tretyakov had been trying to measure the mass of the neutrino since the mid-1970s. In 1980 they announced their exciting results: yes, neutrinos have mass, albeit very little. According to their experiments, neutrinos weigh in somewhere between 14 and 46 eV—about 17,000 times less than electrons.

That's an incredibly tiny mass, but given the incredibly large number of relic neutrinos in the universe, it would be exactly the right amount to solve the dark matter mystery. The invisible stuff in galactic halos; the mysterious matter that prevents galaxies and clusters from flying apart—it might just be our familiar friend, the little neutrino!

Zeldovich was understandably happy. At a banquet speech during an April 1981 conference in Tallinn, Estonia, where the new results were discussed, he said, "Observers work hard in sleepless nights to collect data; theorists interpret observations, are often in error, correct their errors and try again; and there are only very rare moments of clarification. Today is one of such rare moments when we have a holy feeling of understanding the secrets of nature."[3]

Unfortunately, the holy feeling didn't last very long. Competing teams in Zurich and at the Los Alamos National Laboratory in New Mexico were unable to confirm Lyubimov and Tretyakov's results. Instead, they concluded that neutrinos are probably massless, and certainly not heavier than 10 eV or so—not enough to double as dark matter particles. There were other problems, too. For one, if aggregating neutrinos formed the initial "seeds," the formation of the large-scale structure of the universe would have been a top-down process. The first structures to form would be the size of superclusters; only much later would they start to fragment into cluster-sized overdensities and, eventually, into individual galaxies. However, around 1980, astronomers already knew that galaxies formed very early in the history of the universe.

So here's the problem that was facing cosmologists and particle physicists at the start of the 1980s: astronomical observations reveal that the universe is much more massive than it looks. Big bang theory tells you that this "missing mass" ("missing light" would have been a better label) cannot be composed of ordinary atomic nuclei. But the only uncharged nonbaryonic particles that we know of—neutrinos—don't fit the bill. Now what?

Enter Joel Primack.[4] A former particle physicist at Harvard, Primack decided to switch topics in the late 1970s. He was going against the advice of most of his peers, who warned that moving into astrophysics would be dangerous for his career. But Primack loved the complexity and, above all, the confusion of the emerging field of astroparticle physics. Here was an opportunity to work on things no one had ever thought of before. The Standard Model of particle physics had all been worked out and confirmed by an avalanche of discoveries enabled by particle colliders; now it was time to do the same thing for cosmology.

Primack went to the University of California at Santa Cruz. He had an office down the hall from astronomer George Blumenthal, who, along with particle physicist Heinz Pagels of Rockefeller University, became collaborators on a new project. In September 1982 the three scientists published a brief paper in *Nature* titled "Galaxy Formation by Dissipationless Particles Heavier than Neutrinos."[5] The argument was quite straightforward: if neutrinos cannot cluster into galaxy-sized clumps because of their low mass and corresponding high velocity, more massive particles might do the trick. Indeed, the authors showed, if dark matter consists of 1 keV–particles (a thousand electronvolt, which is still just 0.2 percent of an electron mass), the first mass concentrations to form would be more or less galaxy-sized, with a typical mass of one trillion solar masses.

Great, so if dark matter consists of neutral, nonbaryonic particles with a mass of just 1,000 eV, the newborn universe would automatically clump together into galaxies. Problem solved. Except for one important detail: we don't know of any neutral, nonbaryonic

Sandra Faber (left), George Blumenthal (center), and Joel Primack (right) at Lick Observatory in 1984. The photo on the wall shows the nearby spiral galaxy M33.

particles with a mass of 1,000 eV. Go check all the particles in the Standard Model—it's just not there. So did Blumenthal, Primack, and Pagels just make up a particle to fit their needs?

Well, yes and no. True, there wasn't a single thread of observational or experimental evidence for the existence of particles with these properties. But Primack and Pagels were well aware of a proposed extension of the Standard Model in which nature has room for a whole range of new particles, including the so-called gravitino, which would have the right properties. This bold idea, described in more detail in the next chapter, is known as supersymmetry. In fact, Primack had worked on supersymmetry since its inception in the early 1970s. (The roaring seventies again!)

In the abstract of their 1982 *Nature* paper, Blumenthal, Primack, and Pagels wrote, "We suggest here that the gravitino dominated universe can produce galaxies by gravitational instability while avoiding several observational difficulties associated with the neutrino dominated universe." That's a polite way of saying: forget about neutrinos; gravitinos can solve all your problems.

That same year, Jim Peebles published a paper in *Astrophysical Journal Letters,* in which he described a universe dominated by even more massive particles—above 1,000 eV.[6] His motivation? The smoothness problem. All across the sky, the cosmic microwave background has the same temperature, at least to a precision of one part in ten thousand. Apparently baryonic matter must have been distributed extremely smoothly at the time it decoupled from the energetic radiation of the big bang, when the universe was approximately 380,000 years old. Yet the current universe is quite lumpy, as Peebles had already concluded on the basis of the very first galaxy distribution maps, including the *One Million Galaxies* map described in chapter 6.

Peebles's solution to the smoothness problem was that relatively massive, "slow-moving" nonbaryonic particles would show hardly any interaction with photons, if any at all. Since these proposed particles wouldn't be coupled to the dense and energetic radiation field in the early universe, as baryons are, they could start to slowly clump together well before the cosmic background radiation got released. The result, we now realize, would be a three-dimensional web of dark matter overdensities more or less the mass of dwarf galaxies. As soon as baryons (atomic nuclei) also could move freely through space, they got attracted by and started to fall into these "dark matter halos," where densities became high enough for the process of star formation to commence. At a later stage, the resulting "proto-galaxies" would merge into ever-larger structures, eventually leading to the formation of majestic spirals like our own Milky Way galaxy and to giant ellipticals.

Peebles's calculations showed that such massive, weakly interacting particles nicely tie the current large-scale structure of the universe to the big bang's afterglow of 13.8 billion years ago. The weakly interacting particles would yield the correct clumpiness in the distribution of galaxies, while keeping temperature fluctuations in the cosmic background radiation below observed limits. Surprisingly, unlike Blumenthal, Primack, and Pagels, Peebles didn't speculate about the identity of his particles at all—they were pretty much hypothetical. Little wonder that later on, when his suggestion was taken ever more seriously, he sometimes thought, "Hey guys, I'm just trying to solve the smoothness problem, and this is the simplest model I can think of that fits the observations. What makes you think this is right?"

Anyway, Peebles's 1982 paper is generally seen as the birth announcement of the cold dark matter theory, where, as a reminder, "cold" is physics parlance for "slow-moving."[7] The time was ripe. Scientists had been grappling with the concept of dark matter for more than a decade (not counting the early explorations of the 1930s), and, like the blind men in the old Hindu fable, they had all been studying different parts of the same bulky elephant. Now, finally, one theory could explain it all, and everyone jumped on it: radio astronomers, particle physicists, galactic dynamicists, cosmologists, nuclear physicists, cosmic cartographers, numerical simulation programmers, science writers, and college teachers.

So what about the 1984 *Nature* paper coauthored by Faber? That may well have been the publication that raised awareness of the cold dark matter theory among the wider scientific community, not in the least part because one of its four authors was the eminent British astrophysicist Martin Rees.

In spring 1983, both Rees and Primack were attending an international, interdisciplinary physics conference at the Courchevel

Moriond ski resort in the French Alps. The conference program left ample room for sporty scientists to strap on their skis and descend the slopes of Les Trois Vallées. Primack had never tried skiing before, but after one day of lessons—and a lot of falls—he concluded it wasn't for him. Rees also didn't ski, so they ended up talking physics and cosmology in one of the resort's fancy bars instead. Before long, the idea for the cold dark matter paper was born, and Primack asked his Santa Cruz colleagues Blumenthal and Faber to help turn it into reality.

The paper, titled "Formation of Galaxies and Large-Scale Structure with Cold Dark Matter," didn't shy away from some pretty fundamental questions, and it promised to provide satisfying answers.[8] "Why are there galaxies," the authors asked right away, "and why do they have the sizes and shapes that we observe?"

> Why are galaxies clustered hierarchically in clusters and super-clusters, separated by enormous voids in which bright galaxies are almost entirely absent? And what is the nature of the invisible mass, or dark matter, that we detect gravitationally round-about galaxies and clusters but cannot see directly in any wavelength of electromagnetic radiation? Of the great mysteries of modern cosmology, these three may now be among the ripest for solution.

Cold dark matter is the most likely answer, they argued. And while Peebles hadn't speculated about the true nature of the enigmatic stuff, Blumenthal, Faber, Primack, and Rees listed a whole range of potential candidates, including axions, photinos, primordial black holes, and quark nuggets. (I'll come back to axions and primordial black holes in later chapters; the others are so speculative that you can forget about them right away.) The authors went into a lot of detail describing the origin of galaxies and their subsequent aggregation into clusters and superclusters, and they even speculated about the formation of dwarf galaxies and globular clusters—the spherical collections of hundreds of thousands of stars that are found swarming around most major galaxies.

Toward the end of the paper, the authors concluded, "We have shown that a universe with [approximately] 10 times as much cold dark matter as baryonic matter provides a remarkably good fit to the observed universe." The cold dark matter picture "seems to be the best model available and merits close scrutiny and testing."

Dark. Cold. Neutral. Invisible. Nonbaryonic. Massive, in the sense that the particles must weigh at least something—after all, they give themselves away through their gravitational influence. Not subject to electromagnetism or to the strong nuclear force. Possibly interacting through the feeble weak nuclear force. Scientists were finally coming to grips with the properties of dark matter. Now the only thing left to do was to identify the culprit.

It almost looked as if the final answer was waiting around the corner.

10

Miraculous WIMPs

It's a quiet, sunny day in the small French village of Saint-Genis-Pouilly, close to the Swiss border and just ten kilometers northwest of Geneva. Children are playing in front of the fancy detached houses on the Allée Madame de Staël, named after the influential eighteenth-century author and political activist. In the distance are the popular ski resorts in the Réserve naturelle nationale de la Haute Chaîne du Jura. All in all, it's a peaceful scene.

But down below, a nuclear Armageddon is underway. Some 60 meters beneath the village is a 4-meter-wide tunnel. It runs below the Gymnase du Lion sports center, crosses Rue de la Faucille, and heads out of town. Beyond Saint-Genis-Pouilly, the tunnel bends north, to complete a full circle with a circumference of 27 kilometers way below the equally peaceful villages of Gex, Versonnex, Ferney-Voltaire, and Meyrin. Two incredibly narrow bundles of protons—the nuclei of hydrogen atoms—are racing in opposite directions through an evacuated beam pipe in the center of the tunnel ring. These charged particles, described as relativistic protons because they are accelerated to 99.999999 percent of the speed of light, clock over 11,000 revolutions per second. The protons are kept in their circular tracks by more than 1,200 huge superconducting magnets, chilled to 1.9 degrees above absolute zero—colder than outer space.

And just a few hundred meters northwest of the playing children is one of the four "war zones" where the two armies of protons clash into each other with energies up to 13 trillion electronvolt (tera-electronvolt, or TeV), producing showers of subatomic debris in the aftermath.

OK, I'll admit: when I visited the area in June 2019, the Large Hadron Collider (LHC) of the European Organization for Nuclear Research (CERN) was shut down for maintenance and upgrades.[1] Good for me, since it gave me the opportunity to actually go underground, into the caverns that house the gigantic particle detectors. But by the time this book is published, the next major LHC experiment, known as Run 3, will be in full swing: relativistic protons will again crash into each other head-on, and scientists will eagerly study the particles created out of the collision energy, obeying Einstein's iconic equation $E = mc^2$.

The Large Hadron Collider, the most powerful particle collider ever built, has been operational since 2008.[2] But CERN is much older, dating back to 1952. Almost forty years ago, in 1983, a team of CERN scientists led by Italian physicist Carlo Rubbia and Dutch accelerator engineer Simon van der Meer used another particle accelerator, the much smaller Super Proton Synchrotron, to discover the W and Z bosons. These are the massive "carrier particles" of the weak nuclear force. More recently, in 2012, the elusive Higgs boson—the particle that gives mass to other particles—was discovered in data from ATLAS and CMS, the two largest detectors on the LHC circumference. (ATLAS is a contrived acronym that stands for A Toroidal LHC ApparatuS; CMS is the Compact Muon Solenoid.) In the years since, CERN experiments have continued to provide evidence for new and exotic hadrons—particles composed of two, three, or even four or five quarks.

CERN's Large Hadron Collider, built into a tunnel in the vicinity of Geneva, Switzerland, forms a ring 27 kilometers in circumference.

Electroweak bosons, the Higgs particle, even tetraquarks—they're all part of the successful Standard Model of particle physics, just like kaons, pions, charmed Xi primes, and double bottom Omegas, to name a few. Granted, you don't come across them that often; after all, our material world is composed of just protons, neutrons, and electrons. But these unfamiliar entities are all members of the same particle zoo. It's just that most of them are extremely short-lived: within a minute fraction of a second, they decay into more familiar particles. However, if there's enough energy available, as in the collision of fast-moving protons, exotic particles can and will be produced every so often, and they leave telltale tracks and fingerprints in the huge particle detectors surrounding the proton collision points.

Which raises the question: If dark matter particles exist at all, could they also be produced in the Large Hadron Collider? The answer, in principle, is yes—either directly from the energy of two colliding protons, or indirectly, as the decay product of some intermediate particle.

Unfortunately, no one has a clue as to how often dark matter particles may be produced in proton collisions, let alone how massive they are, so no one knows what to expect. Moreover, dark matter particles themselves don't decay into other particles: if they weren't inherently stable, they couldn't possibly constitute most of the mass of the universe! And since dark matter particles hardly interact with "normal" matter, they're almost impossible to spot. In fact, the only way to proceed is to study the results of LHC proton collisions in as much detail as possible and meticulously check the bookkeeping—including the predicted number of newly produced neutrinos, which also go undetected. If the energy or momentum numbers don't add up, something's apparently missing, and that something might well be dark matter.

So far detectors like ATLAS and CMS haven't found a single convincing trace of dark matter. But particle physicists don't give up easily, and in the case of dark matter, they believe they have every reason to keep searching. That's because the detection of weakly interacting massive particles (WIMPs—physicists are extremely fond of acronyms) might not only solve the dark matter mystery but also point the way to exciting physics beyond the Standard Model. In particular, the existence of WIMPs could validate a popular theoretical framework known as supersymmetry.[3]

I briefly mentioned supersymmetry in the previous chapter, and you may have wondered why physicists want to extend the Standard Model if it is as complete and successful as I've asserted. But supersymmetry was first proposed in 1971, when the term "Standard Model" hadn't even been coined yet. Our comprehensive theory of elementary particles and fundamental forces of nature gained general acceptance only in 1983, with the discovery of the W and Z bosons, which turned out to have exactly the properties the Standard Model predicted. And even as scientists have embraced the language of a Standard Model, they have been aware that their existing mathematical description of the physical world can't be the final answer. After all, the Standard Model doesn't account for dark matter or for the tiny mass of neutrinos, which, according to the theory, should be massless. These are just two of the most salient problems.

Anyway, the idea of supersymmetry (affectionately known as SUSY) was almost simultaneously, and largely independently, introduced in the first half of the 1970s by four teams of two physicists each.[4] They all wondered about the puzzling fact that elementary particles come in two populations: fermions (particles of matter, that is, quarks, electrons, and neutrinos) and bosons (force-carrying particles). Might there be some overarching symmetry in nature that would link the two populations together in one description? In that case, fermions and bosons would really be two sides of the same supersymmetric coin. For every known fermion, there would be a corresponding bosonic partner and vice versa.

If you're not a particle physicist, it all sounds a bit ad hoc. But this is how physics has often proceeded: looking for patterns, conjuring up some underlying organizing principle, and predicting new findings on the basis of your theory. It's how Dmitri Mendeleev came up with the idea of the periodic table of elements in 1869. And long before scientists began to understand the composite structure of atoms, he was able to predict the existence of chemical elements that were not yet known. It's also how quantum chromodynamics—the theory of the strong nuclear force—came about:

American physicist Murray Gell-Mann and his PhD student George Zweig discovered a tantalizing mathematical pattern in the properties of subatomic particles, leading them to propose the existence of quarks. Quarks were experimentally confirmed four years later, in 1968.

The nice thing about SUSY is that it not only provides a natural link between fermions and bosons but also solves a number of nagging issues in particle physics. Since this is not a book about particle physics, let alone about supersymmetry, I won't go into all the details. But for one, supersymmetry paves the way to a grand unified theory. The electromagnetic and weak nuclear forces can be described using a single theory, as Sheldon Glashow, Abdus Salam, and Steven Weinberg showed in the 1960s. But the strong nuclear force eluded them. SUSY may provide the tools to incorporate all these forces under a unified theory. SUSY is even a necessary ingredient of string theory—a promising, albeit highly speculative and hypothetical, theory of quantum gravity. Supersymmetry also gives a natural explanation for the fact that the Higgs particle weighs somewhere between 100 and 150 billion eV. Without SUSY, the Higgs could have been much more massive.

Finally, SUSY appeals to experimentalists because it predicts new physics that should occur at collision energies well above the current limit of what can be produced, which is 13 TeV. This limit is important because energy is equivalent to mass and vice versa: more energetic collisions produce more massive particles. As scientists have striven to detect ever more massive particles, they have increased the power of their machines, up to 13 TeV. For now, however, the heaviest elementary particle researchers have detected is the top quark, discovered in 1995 with a mass of "just" 173 billion eV. Nothing has been discovered at energies between that figure and today's limit. If SUSY is right, experimentalists need to keep pushing the limit upward, and eventually new particles will emerge.

CERN is one of the outfits that's extending our experimental reach. Its main site, close to the Geneva Airport, is a sprawling campus of offices, hangars, warehouses, and high-tech laboratories, crisscrossed by a network of roads named after famous physicists, like Route Marie Curie, Route Feynman, and Square Galileo Galilei. In this scientific nirvana, thousands of researchers from all over the world join forces to unravel the most fundamental secrets of nature.

The ATLAS building is adorned with a three-story high cutaway drawing of the detector. When I take the elevator down to the tunnel level, I'm blown away by the sheer size of the instrument: ATLAS is almost half as big as the Notre Dame cathedral in Paris, and weighs as much as the Eiffel Tower. It's so incredibly large that I almost fail to notice the tiny technicians installing new equipment in the detector's innards.[5]

ATLAS is where the first hints of the Higgs boson were found, back in 2012. Here, physicists hope to find evidence for supersymmetry. This is also where dark matter might someday be created and tracked down. For that's yet another bonus of supersymmetry—one that the theory's inventors never thought about back in the 1970s: one of the SUSY particles could very well be the stable WIMP that constitutes the bulk of our universe.

Here's why. Recall that, according to SUSY, every known elementary particle has a supersymmetric partner. All these SUSY particles must be more massive than the "normal" particles we know of, otherwise they would have been produced and detected in collider experiments already. Moreover, much like most of the particles in the Standard Model, SUSY particles are expected to be unstable and to decay into lighter particles, including Standard Model members.

But there's a catch. In many viable versions of SUSY, if a supersymmetric particle decays, at least one of the less massive decay products has to be supersymmetric, too. For complicated reasons, if this weren't the case, our good old friend the proton would be unstable, falling apart within a year, or maybe even within a fraction of a second. Fortunately for us, protons are as stable as anything, so we must assume that, indeed, SUSY particles can't decay into just Standard Model particles and nothing else.

But that means that the lightest supersymmetric particle, also known as the LSP, has to be stable! According to theory, the lightest supersymmetric particle is a so-called neutralino, which, as the name implies, has no electrical charge. It also doesn't feel the strong nuclear force. Stable, neutral, massive, and subject to the weak force—there you go: the LSP could very well be the WIMP that constitutes the dark matter in the universe.

As we saw in the previous chapter, in the early 1980s astronomers and cosmologists had concluded that their "missing mass," as some were still calling it then, most likely would consist of relatively slow-moving particles—cold dark matter. One of the candidate particles was the hypothetical axion, which I'll get back to in chapter 23. Despite their extremely tiny mass, axions are slow movers, so they were considered a possible constituent of cold dark matter. But before long, the much more massive WIMP—and especially its SUSY version—became everyone's favorite dark matter candidate. Who knows, supersymmetry—a single, promising extension to the Standard Model—might lead the way to a grand unified theory and solve the nagging dark matter riddle at the same time.

And then there was the WIMP miracle. Normally scientists don't believe in miracles, but this one seemed too good to ignore.

To understand the WIMP miracle, you need to recall that the very early universe was a seething cauldron of high-energy photons and short-lived particles—a roiling brew of energy and mass. Right after the big bang, $E = mc^2$ was all over the place, as was $m = E / c^2$. In other words, particle-antiparticle pairs were continuously created out of pure energy, and a split second after they emerged, they annihilated each other, with matter turning back into radiation only to birth new matter again.

But as the newborn universe cooled down, the photons became less and less energetic. As a result, the spontaneous production of the heaviest particle pairs came to a halt. Meanwhile, the expansion of the early universe was rapidly diluting the particles and corresponding antiparticles that had been formed at an earlier stage, so they didn't encounter each other as frequently as they used to. Mutual annihilation of heavy particles and their corresponding antiparticles could still occur, but a fraction of the initial inventory survived the onslaught.

Using the equations of the big bang—basically an expanding and cooling gas, so high school physics will do—it's relatively straightforward to calculate what the remaining "relic density" is for a particular type of particle. And if you do the math for WIMPs, which probably are their own antiparticles, it turns out that you arrive at almost exactly the density astrophysicists and cosmologists have deduced for their cold dark matter. Pretty miraculous, right?

Since they interact through the weak nuclear force (as well as through gravity, of course), WIMPs are expected to have masses on the order of a few hundred thousand eV—a few hundred times more massive than protons. But for the WIMP miracle to work, the precise value of the mass isn't that important. If they're more massive, pair production stops at an earlier stage, when the density of the young universe is still very high. As a result, mutual annihilation is more efficient, and fewer relic particles remain. In contrast, if WIMPs weigh less, pair production can continue for a longer time, and when it finally stops, the cosmic density is lower, with more particles escaping annihilation. But the end result—fewer massive particles or more lightweight ones—always gives you more or less the same average mass density, and it's surprisingly close to the value that people like Jim Peebles, Sandra Faber, and Joel Primack had found for the amount of dark matter in the universe.

Let's pause for a recap. Galaxy dynamics tells you there must be more matter in the universe than meets the eye. Big bang nucleosynthesis reveals that the matter can't all be baryonic. Moreover, nonbaryonic matter can explain the clumpiness of the universe without contradicting the smoothness of the cosmic microwave background. With all this in mind, it makes sense for physicists to go looking for nonbaryonic particles, but these must also be electrically neutral, for, if dark matter comprised charged particles, they would be easy to find. The only neutral nonbaryonic particles that we know of are neutrinos. But neutrinos are not massive enough to double as dark matter particles, so we need to look for unknown types of matter. What then? Fast-moving particles don't clump on the right scales to explain the early formation of galaxies, so dark matter must be cold instead. Weakly interacting massive particles fit the bill perfectly— if they exist, they are expected to yield exactly the right mass density. And supersymmetry predicts the existence of one particular WIMP: the lightest supersymmetric particle, also known as the neutralino.

In the mid-1980s, it was evident what the next step should be: find the damned thing. Particle physicists had found the tau (or tauon, a short-lived, heavy cousin of the electron) in 1975 at the Stanford Linear Accelerator Center (now the SLAC National Accelerator Laboratory) and the W and Z bosons in 1983 at CERN's

Super Proton Synchrotron. A more powerful machine would surely succeed in discovering WIMPs, and confirming supersymmetry at the same time, wouldn't it?

Unfortunately, it didn't work out that way. Nature isn't always kind, as Peebles has been known to say.

The vast majority of the physicists I meet at CERN were in kindergarten or not born yet in the mid-1980s, when their older colleagues dug a 27-kilometer tunnel and built the Large Electron-Positron collider to go searching for mystery particles. But no WIMPS were found. Then came the much more powerful LHC, and still researchers have never hit the WIMP jackpot or found any evidence for SUSY. Discovering the Higgs particle was great of course, and it's certainly exciting to learn more about weird particles like pentaquarks or about the quark-gluon plasma that may have filled the very early universe. But, at this point, after decades of effort, no physics has been encountered beyond the Standard Model, leading to frustration.

During my visit, I meet John Ellis in his surprisingly small office at CERN's theoretical physics department. Ellis, who has been at the European lab since the 1970s, has been a vocal proponent of supersymmetry since its inception.[6] Back in 1984 he was among the first to show how the lightest SUSY particle could be a candidate for dark matter, in a *Nuclear Physics B* paper written with John Hagelin, Dimitri Nanopoulos, Keith Olive, and Mark Srednicki.[7] Although nothing has been found so far, Ellis still believes that WIMPs are more promising as dark matter particles than axions are. So what's his take on the lack of experimental evidence?

"It just tells me we need to look harder," Ellis replies. "WIMPs may be more massive than we have assumed." The problem with that, he adds, is that the WIMP miracle falls by the wayside if the particles are too heavy. "The options are limited. At a mass of around

10 TeV"—ten thousand times the mass of a proton—"you run out of wiggle room. But to probe that mass range, we would need an even bigger detector than the LHC. I don't know when we will find the answer."

"When." Ellis doesn't say "if."

Twenty-six years elapsed between the prediction of the neutrino, in 1930, and its discovery, in 1956. For the Higgs boson, the wait lasted forty-eight years. Gravitational waves—minute ripples in the very fabric of spacetime—were predicted by Albert Einstein in 1916 and weren't discovered until almost a century later, in 2015. Yes, the dark matter hunt at CERN is taking longer than we had expected. But absence of evidence isn't evidence of absence. Who knows what a next-generation collider will reveal. Who knows what Run 3 will reveal.

Maybe that's the real problem with dark matter. We don't know exactly what we're looking for, so there's always good reason to keep on searching. Think of a terrestrial treasure hunt. If you knew the exact location of some mythical city, you could just go there and explore the area. If you didn't find the city, you would conclude that it's just a myth and call off the search. But if you're sailing the seven seas in search of a magical island that could be anywhere on the globe, you shouldn't stop your quest just because you feel it's taking too long. For all we know, the island could be just beyond the horizon.

The discovery of WIMPs may also be just beyond the horizon. Only time will tell. Time, ingenuity, and perseverance.

I I

Simulating the Universe

In the beginning, the universe is without form and void, and darkness is over the face of the deep.

Then I witness how minute density variations in the distribution of dark matter particles begin to evolve into a three-dimensional cobweb-like pattern. Hydrogen and helium atoms—more familiar but much less numerous—follow suit; they can't help being pulled into the same large-scale structures by the sheer gravity of the weird, invisible stuff.

All around me, I now see gas streaming along serpentine filaments, ending up in the high-density regions where these cosmic tentacles meet. Swept up by gravity and twisted by pervasive magnetic fields, the clouds of gas become much more turbulent than the invisible substrate of dark matter on which they are condensing. As hundreds of millions of years pass by in mere seconds, gas starts to collect in the cores of more-or-less spherical halos of invisible dark matter. Slowly but surely, the universe gives birth to a small cluster of galaxies.

In the distance, in the cluster's core, I see how puny dwarf galaxies—traces of dark matter clumps—collide and merge into an ever-growing whole. Meanwhile, right before my eyes, a huge gas cloud further collapses under its own weight and starts to spin faster and faster, slowly flattening in the process. Gobbling up smaller satellite systems, it evolves into a beautiful spiral galaxy.

To my right, two spirals crash into each other, slinging away tidal tails of galactic debris. Shocks and density waves produce a baby boom of new, massive stars. Eventually, the resulting merger settles down as a huge elliptical galaxy, surrounded by concentric shells of gas. To my left, the further growth of yet another disk galaxy is inhibited by energetic supernova explosions in its spiral arms and by powerful outflows from its core, where a supermassive black hole is feasting on infalling gas and blowing some of it back into space.

Zooming in on the relatively quiet spiral galaxy in front of me, I can't wait for the passage of the next nine billion years of accelerated cosmic time. That's when a yellow, run-of-the-mill star will be born out of a small cloud of gas and dust, somewhere on the inner edge of one of the spiral arms. Orbiting the inconspicuous star will be a tiny, rocky planet—a mote of dust in the cosmic ocean. Before long, hydrocarbons raining down from outer space will turn this barren place into a fertile world, brimming with life. Billion-year-old carbon.

But that's only happening in my imagination, for I'm not watching the evolution of the real universe. I've lost myself in video footage of a highly detailed three-dimensional computer simulation called IllustrisTNG (The Next Generation).[1]

IllustrisTNG doesn't simulate the origin of life, but it's still pretty impressive. Fourteen billion years of cosmic evolution, structure formation in an expanding universe, spiral galaxies with dark matter halos—it's all there, and it looks uncannily realistic. It's hard to shake the impression that this is just a sped-up version of the real universe. Like a prosecutor standing before a jury and minutely reconstructing a crime, the simulation so convincing that you can't help thinking it must have happened this way.

Still image from an IllustrisTNG computer simulation of the growth of the large-scale structure of the universe.

Today computer simulations are an indispensable part of the astrophysicist's toolkit. Some forty years ago, however, things were different. Physics—and astrophysics—was very much analytical, and progress was usually made by algebraically solving intricate polynomial or differential equations. In fact, Stephen Hawking once remarked that using a computer to solve a problem in general relativity would destroy the beauty of physics.

So when four young, audacious astronomers started to simulate the whole universe on their computers in the early 1980s—an effort that eventually yielded far-reaching conclusions about the possible nature of dark matter—it wasn't so surprising that they met with skepticism. Indeed, they became known as the Gang of Four, after the group of radical Chinese Communist Party officials who were influential during the last stages of Mao's Cultural Revolution. But while their peers were suspicious and reluctant at first, Marc

Davis, George Efstathiou, Carlos Frenk, and Simon White are now seen as bold pioneers.[2] Their numerical simulations of the evolution of large-scale structure in the universe are the basis of present-day projects like IllustrisTNG.

How do you simulate a universe? Or, more precisely, how do you simulate structure formation in the universe? It's really not that complicated. The Gang of Four focused on nonbaryonic dark matter (the major material constituent of the universe), which doesn't emit or absorb any light, doesn't heat up or cool down, and doesn't respond to magnetic fields. The only game in town is gravity, so you can use the same approach Jim Peebles and Jerry Ostriker did when they ran computer simulations of the evolution and the stability of disk galaxies (see chapter 4). It's all about an initial distribution of test particles, each one representing a certain amount of dark matter. The computer code calculates the mutual gravitational attraction of the test particles in incremental time steps. Here, too, more test particles and smaller time steps increase the reliability of your results. This sort of system is called an N-body simulation: a model of how a large number of objects (in this case, quantities of dark matter particles) interact under the influence of their mutual gravity.

You can't do this for the entire universe, of course. Instead, you just consider a large enough cubic chunk of expanding space, assuming that it is representative of the universe as a whole. "Expanding" is a key word here: with every time step, your cubic chunk of space will grow a little bit, the distances between your test particles will increase, and their mutual gravitational attraction will get a bit weaker. Eventually, the formation of large-scale structures is the result of a tug-of-war between gravity and cosmic expansion.

The smoothness of the initial distribution of test particles is critical. If the distribution were perfectly smooth, not much would happen in your expanding chunk of space, so you need to start with tiny density fluctuations. Areas with a slightly higher-than-average dark matter density will spread out and dilute over time thanks to the expansion of the universe, but they will do so more slowly than areas with a lower-than-average density. The end result is that the relative density variations tend to increase—the contrast between over-dense and under-dense areas strengthens as the eons go by.

Finally, simulations need to take into account what type of dark matter we're talking about. As we saw earlier, there's a big difference in behavior between hot (fast-moving) particles like neutrinos and cold (relatively slow-moving) particles like WIMPs: hot particles can cluster only on very large scales, while cold particles will aggregate into smaller clumps.

The resulting dark matter distribution will determine where galaxies will form, since the less abundant baryonic matter in the universe (basically atomic nuclei) is expected to flow toward the regions with the highest nonbaryonic matter density. In other words: galaxies are expected to form where dark matter is clumped most strongly.

So in the end, a number of assumptions—or, if you will, initial conditions—go into your simulation of the universe: the overall matter density, the type of dark matter (hot or cold), the spectrum of initial density fluctuations, the cosmic expansion rate, and so on. But once you've set all of the dials at the desired values, you just push the start button and wait to see what kind of universe this particular choice of initial conditions will produce after billions of years of evolution.

In the late 1970s, Marc Davis, the oldest member of the Gang of Four, already knew what kind of universe should be produced. In 1977, at Harvard, Davis had started the Center for Astrophysics (CfA) redshift survey, together with John Huchra, David Latham, and John Tonry (see chapter 6). Their first rudimentary 3D map of the distribution of galaxies in the "local" universe wasn't published until 1983, but initial results had shown that galaxies were grouped in giant walls and filaments, surrounding relatively empty voids. Any credible theory—or computer simulation—of the universe should at least be able to explain or produce this specific kind of large-scale structure.

Most astrophysical N-body simulations at the time were restricted to something like a thousand test particles or so.[3] In a 3D simulation, that amounts to a cube of just ten by ten by ten particles—way below the number you need to simulate a universe. But in 1979, Davis learned about novel computer code that could do much better. He was on his way to an international cosmology conference in Tallinn, Estonia, and the easiest way to get there was by ferry across the Baltic Sea from Helsinki, Finland. On board, he met George Efstathiou, who was traveling to the same conference. Efstathiou was a young British grad student, the child of Cypriot immigrants. He had no money; Davis bought him supper, and they became friends for the rest of their lives.

Efstathiou had been in touch with condensed matter physicists who were studying melting processes in atom lattices. Very different from cosmology (for one, gravity doesn't play any role on the scale of atoms), but those scientists had developed computer code that could handle cubes of $32 \times 32 \times 32$ elements—a whopping 32,768 test particles! Efstathiou was busy converting this code into something that could be used for cosmological purposes. Maybe that would finally enable simulations detailed enough to compare to the CfA redshift surveys—the only available 3D map of the real universe at the time.

Davis had met another member of the gang earlier, during a sabbatical at the University of Cambridge. Simon White had started out as an applied mathematics graduate student, studying in a stuffy, windowless basement in a downtown university building. But after visiting Cambridge's Institute of Astronomy just west of town, with its sunlit rooms and daffodil-lined lawns, he decided to switch fields. The two met again at the University of California, Berkeley, where White became a senior fellow in 1980 and Davis secured a tenured position in 1981. By then, combining math and astronomy, White was developing computer code to simulate gravitational interactions in galaxy clusters. Would he be interested in an attempt to simulate the whole universe? You bet!

Meanwhile, back in England, Efstathiou had become friends with grad student Carlos Frenk, the son of a German-Mexican medical doctor and a musician. After earning his PhD in 1981 with White in Cambridge, Frenk went to Berkeley to become one of Davis's first postdocs, working on the analysis of the CfA redshift survey results. And Efstathiou, who had held a postdoc position at Berkeley before but had returned to Cambridge, would regularly fly to California to join his friends and help realize the ambitious goal of simulating the growth of structure in the universe.

Back then, powerful computers were large and slow and scarce. The Berkeley machine—a Digital Equipment VAX-11 / 780—filled the better portion of a room, but it ran on a mere 16 megabytes (MB) of internal memory. One simulation easily took more than a full day. For comparison, a current off-the-shelf MacBook would be able to complete the task in fewer than 30 seconds.

Taking advantage of the Starlink computer network—mutually connected VAX computers at astronomical research centers throughout the United Kingdom—Efstathiou and Frenk used every machine they could lay their hands on. When it turned out that you could only use Starlink for a maximum of two hours at once, after which you had to apply for more computer time, Efstathiou wrote a script that cleverly overruled this limitation. Sure, other researchers complained that they couldn't get access to the network, but what could be more important than simulating the evolution of the universe?

The first simulations, published in 1983 by White, Frenk, and Davis, showed that hot dark matter (neutrinos, for instance) could not reproduce the real universe.[4] Fast-moving particles were shown to slowly cluster together in very large structures, comparable in size to superclusters of galaxies. These structures need to fragment into smaller clumps before galaxies can form. And because of this top-down scenario, the smallest matter concentrations—the seeds of galaxies—are only found within the supercluster-sized structures. The voids in between the superclusters remain completely empty in the simulations.

In contrast, observations reveal that galaxies formed quite early in the history of the universe, before the formation of superclusters. Moreover, voids are not completely empty; they, too, contain iso-lated galaxies, albeit in small quantities. This is exactly the outcome of simulations with cold dark matter, which soon became the team's sole focus. Because of the lower particle velocities, cold dark matter first clumps into small dark matter halos, more or less the size of dwarf galaxies. Once the first small galaxies have formed (by accre-tion of baryonic matter), most of them will start to merge into larger galaxies, which successively gather into groups, clusters and, even-tually, superclusters—a process that is still very much ongoing in the universe.

The Gang of Four—the nickname was coined by Berkeley as-trophysicist Chris McKee—worked feverishly over the Christmas holidays in late 1983 and during a four-month workshop on the large-scale structure of the cosmos in Santa Barbara in 1984. In May 1985, they published their first results and conclusions in *The Astrophysical Journal*.[5] The title says it all: "The Evolution of Large-

Scale Structure in a Universe Dominated by Cold Dark Matter." "It is remarkable how many aspects of the observed galaxy distribution are reflected quite faithfully by the distribution of CDM," the authors write. "This seems too good to be true, but perhaps it hints that we are at last approaching a correct resolution of the missing mass problem."

In a shorter follow-up paper in *Nature,* published in October, the group presented simulations that showed the formation and occasional merging of individual dark matter subhalos, leading to a pretty realistic population of disk galaxies (with flat rotation curves and all) and ellipticals.[6] Was cold dark matter really solving all the riddles astronomers had been struggling with? It sure looked like that. The fact that no cold dark matter particle had ever been observed suddenly seemed to be a minor detail. "These people are magicians," commented Princeton astrophysicist Richard Gott, who was a referee on the *Nature* paper.

And the magicians weren't done yet. In 1987 and 1988, they published three more papers—two in *The Astrophysical Journal* and one in *Nature*—in which they expanded on their earlier work.[7] Taken together, the Gang's five landmark publications—collectively known as the DEFW papers, for Davis, Efstathiou, Frenk, and White—firmly put nonbaryonic cold dark matter on the map as the sole candidate for the major constituent of the universe. CDM appeared to be able to explain just about everything.

A major question remained, though: How much dark matter does the universe contain? In the vast majority of their initial simulations, the computer wizards had assumed an overall mass density of the universe equal to the critical density—the amount of gravitating matter that would eventually bring cosmic expansion to a halt without reversing into a collapse. Since the baryonic matter produced by big bang nucleosynthesis accounts for only 5 percent of the critical density, nonbaryonic cold dark matter would have to constitute the remaining 95 percent—a staggering imbalance, much more than one would infer from galaxy dynamics.

The four astronomers came to realize that a critical-density universe was basically just an "aesthetically pleasing idea," as they called it. (We'll get back to this in chapter 15.) Nature of course has no compelling reason to meet human aesthetic needs, so what if the total mass density of the universe were much *lower* than the critical value, and more in line with the earlier mass estimates by Ostriker, Peebles, and Yahil; Gott, Gunn, Schramm, and Tinsley; and Faber and Gallagher?

Indeed, White and Frenk, together with Julio Navarro and August Evrard, concluded in 1993 that either we do not understand big bang nucleosynthesis or the universe cannot have the critical density. The argumentation in their *Nature* article is quite straightforward.[8] Returning to the Coma Cluster of galaxies (the subject of Fritz Zwicky's much-ignored 1933 paper), they first derived the total dynamical mass of the cluster from the velocities of its member galaxies—the same method that Zwicky had applied. Next, they determined the baryonic mass, taking into account not just the visible galaxies—stars and nebulae—but also the huge amounts of extremely hot gas that X-ray telescopes had revealed between the cluster galaxies. By comparing the two mass estimates, the authors found that the baryonic mass in the Coma Cluster makes up about one-sixth of the total gravitating mass. For other clusters, similar values were found.

But if baryons account for only 5 percent of the critical density— that's what big bang nucleosynthesis tells us—and if the universe indeed has the critical density, then the nonbaryonic dark matter in the universe would have to be nineteen times as abundant as the baryonic matter in the form of "normal" atoms, not six times. And given the way in which galaxy clusters are thought to form in an expanding universe (based on the type of computer simulations that the Gang of Four had pioneered), there's just no way they can end up with a baryon fraction that's more than three times higher than the average cosmic value. In other words, the high baryon fraction of clusters like Coma must reflect the cosmic average, in which case the universe cannot have the critical density.

With the discovery of the accelerating expansion of the universe in 1998—attributed to yet another mysterious cosmic component called dark energy, described in chapter 15—it became clear that the total mass density of the universe is much lower than the critical density: something on the order of 27 percent of the critical density. Ever since, computer simulations of the growth of cosmic structure have used this value for the amount of gravitating matter in the universe and taken dark energy into account, too. Thanks to an incredible increase in computing power, these simulations are of course much more detailed than those of the Gang of Four, and the close correspondence between contemporary simulations and the real universe out there has contributed considerably to the general acceptance of what is now known as the cosmological concordance model.

To give you an idea of the progress that has been made since the early 1980s, let's take a look at the groundbreaking Millennium Simulation, run in 2005 by members of the so-called Virgo Consortium.[9] Also known as the Millennium Run, the project was led by Volker Springel of the Max Planck Institute for Astrophysics in Garching, Germany; the first *Nature* paper on the Millennium Simulation results was coauthored by White (Springel's thesis advisor), Frenk, Navarro, and Evrard, among others.[10] While White and Frenk had started out with simulations of just 32,768 particles in a $32 \times 32 \times 32$ cube, the Millennium Simulation followed the mutual gravitational attraction of no fewer than 10 billion dark matter test particles (2160×2160×2160). And instead of a VAX computer with just 16 MB of internal memory, Springel and his colleagues used an IBM Regatta supercomputer with one terabyte of memory (1 TB, approximately one million MB). Performing 200 billion floating point operations per second, it took this monster machine 28 days—a total of 343,000 processor hours—to complete the simulation, yielding 27 TB of stored data, all of which has been made available to the scientific community.

Like the original Gang of Four simulations, the Millennium Run dealt only with the clumping of dark matter, which is relatively easy because you only need to consider gravity. But what about the baryonic matter? How do familiar atoms aggregate on the invisible skeleton of nonbaryonic dark matter? How do real galaxies form? That's a much more complicated question, since atomic nuclei (and electrons) are governed not only by gravity but also by radiation, collisional gas drag, and magnetohydrodynamic processes, to name just a few nasty examples. Moreover, since baryonic matter interacts with light, it can heat up and cool down by absorbing or emitting energy.

Astronomers have recently succeeded in developing humongous computer simulations that take all these complications into account. By using a broad range of mathematical tricks, they are now able to model messy problems like cooling flows, galactic winds due to the explosions of massive stars (supernovae), and the energetic effects of supermassive black holes in the cores of galaxies.

In late 2014 and early 2015, two competing groups published results from such enriched simulations, accounting for nonbaryonic and baryonic matter alike. The models, Illustris and EAGLE (Evolution and Assembly of GaLaxies and their Environments), both take you on a mind-boggling tour through space and time, from the very first density perturbations in the early universe all the way to the formation of irregular dwarf galaxies, majestic spirals, and bulky el-lipticals.[11] As of this writing, the state of the art is a new version of Illustris, the 2017 IllustrisTNG simulation, which can follow the be-havior of more than 30 billion test particles (both dark matter and gas) in a cubic chunk of space that expands to a current size of al-most one billion light-years across.

Upon publication of the EAGLE simulation in a 2015 paper in *Monthly Notices of the Royal Astronomical Society,* coauthor Richard Bower of Durham University said, "The universe generated by the computer is just like the real thing. There are galaxies everywhere, with all the shapes, sizes, and colors I've seen with the world's largest telescopes. It is incredible."[12] And the end is not yet in sight, says EAGLE project lead Joop Schaye of Leiden University—in principle, you could go on forever, looking in ever more detail at the birth of stars and the formation of planets.

No one expects that we will be able to simulate the origin of life any time soon. However, flying through eons and gigaparsecs in the IllustrisTNG simulation, witnessing how small variations in the den-sity of dark matter evolve into the large-scale structure of the uni-verse, and zooming in on a budding spiral galaxy in the outskirts of a populous cluster gives you a unique perspective on your own place in time and space. This is how it may have happened. Almost 14 billion years after the big bang, on a grain of sand orbiting a pinprick of light, a curious species started to think about its cosmic roots and its miraculous connection to the great wide open.

Without the prodigious amounts of cold dark matter that fill our universe, we would probably not be here. And even though we have no clue yet as to the real nature of dark matter, we can now be ab-solutely sure that this mysterious stuff is at the very basis of our existence.

Or can we?

12

The Heretics

I have always felt sympathy for scientific rebels. People who choose to swim against the tide. "Everybody says X? Well I believe it's Y." These are creative characters, not easily discouraged by fierce opposition or even ridicule. And no, I don't mean pseudoscientists who claim that the pyramids were built by aliens or crackpots working on a perpetuum mobile. I'm talking about real scholars, questioning or even attacking prevailing wisdom with original thinking and solid arguments. Iconoclasts.

So when, as a teenager, I read my first astronomy books by the Dutch teacher and science writer Tjomme de Vries, I loved the story about Fred Hoyle and his steady-state model, which took issue with the conventional-wisdom big bang theory of the origin of the universe. And in the mid-1980s, as a beginning science journalist, I got intrigued by the theories of Halton Arp and Margaret Burbidge, who argued that galaxies and quasars might not be as remote as you would infer from their redshifts. What if those dissenters were right?

It must not have been much later that I came across the work of Israeli physicist Mordehai Milgrom—probably in the 1988 book *The Dark Matter* by Wallace and Karen Tucker.[1] Here was someone with a fresh take on a nagging cosmic mystery. While astronomers were becoming convinced that the flat rotation curves of galaxies and the dynamics of galaxy clusters could only be explained by assuming that the universe is dominated by dark matter, "Milgrom took another approach," the Tuckers wrote. "He tried to change the laws of physics." Now here was some heretic.

Come to think of it, Milgrom's idea makes a lot of sense. The velocities of galaxies in clusters are much too high. The outer parts of disk galaxies are rotating much too fast. Galaxies and groups of galaxies are much too heavy. Sure, there's nothing wrong with the measurements themselves. But what about the qualifications too high, too fast, and too heavy? Those are all based on our assumption that we understand how gravity works. Yet if gravity behaves differently on cosmic scales, then everything might actually be just fine. We wouldn't need dark matter at all to explain our observations.

If Milgrom were right, it wouldn't be the first time that an observational riddle was solved by tweaking our theory of gravity. It happened just over a century ago.

In the first half of the nineteenth century, astronomers had noted how Uranus was diverting from its predicted path. Apparently something was tugging on the remote planet. French mathematician Urbain Le Verrier used Isaac Newton's law of universal gravitation to calculate where the culprit might hide, and, sure enough, Neptune was found in 1846 near the predicted position.[2]

But the solar system's innermost planet, Mercury, was slightly misbehaving too. Encouraged by his earlier success, Le Verrier tried to pull off the same mathematical trick again, and in 1859 he proposed the existence of an "intra-Mercurial" planet, which was called Vulcan. But Vulcan was never found, and we now know it doesn't exist (at least not outside the universe of *Star Trek*). Instead, Mercury's "errant" behavior was fully explained by Albert Einstein's 1915 general theory of relativity—an improved version of Newton's formulation of gravity.[3]

What else might be wrong—or at least incomplete—with our understanding of gravity? For instance, we all learned in high school that the pull between two massive bodies decreases with the square of the distance between them. Sensitive laboratory experiments and observations within our solar system confirm this so-called inverse square law. But how can we be so sure that it holds throughout the universe?

In his 1937 paper on the Coma Cluster, Fritz Zwicky was careful enough to note that his conclusions about the mass of the cluster rested "on the assumption that Newton's inverse square law accurately describes the gravitational interactions among [galaxies]." Likewise, Horace Babcock, in his 1939 thesis on Andromeda, concluded that there must be large amounts of dark mass in the outer parts of the galaxy, "or, perhaps, that new dynamical considerations are required"—in other words, novel ways to treat gravity. Italian astrophysicist Arrigo Finzi went one step further in 1963, by actually proposing how gravity might work differently on very large scales.[4]

It would take twenty more years before Mordehai Milgrom published his theory of modified Newtonian dynamics, known as MOND. If correct, the theory would undermine the necessity of dark matter. As of this writing, the jury is still out. Some revolutions proceed extremely slowly; many never happen at all.

In September 2019 I met Milgrom at a five-day workshop in Bonn, Germany.[5] A tall, slim person, casually dressed in a black T-shirt, black trousers, and sneakers, he was sitting in the front row at every talk, asking questions and initiating lively discussions. In between presentations, he took considerable time to tell me his story.

Trained as a particle physicist, Milgrom, who goes by Moti, has been affiliated with the Weizmann Institute of Science in Rehovot, Israel, since the 1970s. In 1980 and 1981, during a sabbatical at the Institute for Advanced Study in Princeton, he delved into the

Mordehai Milgrom (left), talking with University of Geneva astrophysicist André Maeder at the Bonn workshop on modified Newtonian dynamics, September 2019.

emerging field of galaxy dynamics and learned about the curious fact that galactic rotation curves always seem to become flat at large distances from the center.

Dark matter, right? That's what everybody says. But what if there's something wrong with Newton's laws? What happens if you assume that flat rotation curves are produced by some non-Newtonian form of gravity? At first Milgrom was pretty skeptical himself. "If you'd asked me back then whether this would ever lead to something useful, I would've given it a slim chance," he says. Surprisingly, though, he didn't run into any theoretical inconsistencies as he tried to make sense of the strange observations. Slowly but surely, it became evident that a simple modification of Newtonian gravity could explain flat rotation curves in one fell swoop.

Back home in Israel, Milgrom feverishly worked out all the details—an obsessed thirty-five-year-old scientist, certain that he was onto something big. "I hardly slept. I kept a notebook next to my bed. My wife tells me I was out of touch most of the time." And he kept everything to himself, lest colleagues call him crazy, or—worse still—steal his ideas. "I was absolutely sure everyone would jump on it; that's how convinced I was," he says.

But when Milgrom privately sent his three papers on modified Newtonian dynamics to five eminent theoretical astrophysicists, including Martin Rees and Jerry Ostriker, none of them was overly enthusiastic—although they also didn't think him completely nuts. And when he submitted the first paper for publication, *Astronomy & Astrophysics, The Astrophysical Journal,* and *Nature* all rejected it. Only after a long and frustrating struggle with the editors did *The Astrophysical Journal* finally accept Milgrom's second and third papers, on the implications of his new idea for galaxies, galaxy groups, and galaxy clusters. After that, he talked the journal into running the first paper, too.

The three papers were eventually published back to back in the July 15, 1983, issue.[6] The very first sentences of the first paper reveal Milgrom's self-confidence, which has never really left him. "I consider the possibility that there is not, in fact, much hidden mass in galaxies and galaxy systems," he wrote. "If a certain modified version of the Newtonian dynamics is used to describe the motion of bodies in a gravitational field (of a galaxy, say), the observational results are reproduced with no need to assume hidden mass in appreciable quantities."

There you have it. Dark matter doesn't exist.

To explain Milgrom's hypothesis, I'll have to bring the concept of a rotation curve back to mind. In our solar system, Neptune has a much lower orbital velocity than Mercury, since it is much farther from the Sun, while gravity falls off with the inverse square of distance—at least, according to Newton. Less gravitational attraction means lower speed. A plot of orbital velocity against distance shows this reduction in speed—a characteristic and continuous curve known as the Keplerian decline, named after Johannes Kepler, who was the first to formulate mathematical laws of planetary motion, in the early seventeenth century.

The rotation curve of a galaxy is expected to look somewhat different from that of a planetary system. To see why, compare our solar system to a galaxy. While almost all of the mass of the solar system is concentrated in the Sun, a galaxy's mass is distributed over a much larger volume. It turns out that the orbital velocity of a star (or any other object) in the galaxy is determined not only by the mass at the galaxy's center but also by the total amount of mass at smaller distances from the center. Still, at larger distances, in the galaxy's dark outskirts, you would expect something close to a Keplerian decline: the more distant a star (or a cloud of hydrogen gas) is from the center of the galaxy, the slower it should orbit.

Instead, as radio observations reveal (see chapter 8), velocities remain constant way beyond the visible disk of a galaxy. In other words, the rotation curve reaches a certain terminal velocity, after which it stays flat, suggesting the existence of large amounts of invisible gravitating matter. This doesn't mean that galaxies rotate the way solid objects like wagon wheels do: distant orbits have a larger circumference, so despite moving at the same velocity as a star nearer to the center of the galaxy, a more distant star takes longer to complete a revolution.

However, the flat rotation curve raises a critical question, which implicates the dark matter theory: Why would dark matter be precisely distributed in such a way as to produce flat rotation curves, as opposed to some other form of slower-than-Keplerian decline? Milgrom's answer is simple. If gravity falls off with the inverse of distance—and not with the inverse of the *square* of the distance—you will automatically end up with a flat rotation curve. No need for dark matter and no need to figure out why it is distributed in a manner that produces flat rotation curves. Problem solved.

But wait, gravity evidently does not behave this way in our solar system. So what sets a galaxy apart from a planetary system? Why would gravity behave differently out there than it does right under our noses? According to MOND, it all has to do with the strength of the gravitational field. If the field strength gets below a certain limit, gravity changes face, and Newton's inverse square law no longer holds. On the surface of the Earth, we experience a convenient gravitational field of 1 g (which equals a gravitational acceleration of 9.81 m / s^2). On the Moon's surface, the field strength is just 0.16 g. Meanwhile, the Moon is kept in its orbit by the Earth's gravity, which, at the Moon's distance, amounts to a mere 1 / 3600 g (that's because the Moon is 60 times farther away from the Earth's center than our planet's surface is). Likewise, it's easy to show that the Sun's gravitational field as experienced by the distant dwarf planet Pluto is just 0.00000067 g.

As far as MOND goes, these are all huge numbers. But things get different in the outer regions of galaxies and in intergalactic space, where field strengths are much, much lower. Milgrom's "tipping point"—where gravity gradually starts to behave differently— lies around one hundred billionth g (corresponding to a gravitational acceleration of 1.2×10^{-10} m / s^2, to be precise). If Earth's gravity were that weak, it would take an apple two days to fall down from a height of one meter.

This may all sound a bit speculative and contrived, and it is. Then again, over the centuries, scientists have always tried to find simple mathematical rules and laws to describe their observations and measurements as best as they could. And modified Newtonian dynamics successfully describes the observed rotational properties of galaxies. "I didn't know of any physical reason why it would work," says Milgrom, "but it worked."

In fact, it worked better than expected. Already in his second *Astrophysical Journal* paper, Milgrom made a testable prediction about the relation between the luminosity of a galaxy and its "terminal velocity"—the rotational velocity in the outer regions. Assuming that a galaxy's energy output is a proxy for its total mass (just gas and stars, according to MOND), it can easily be shown that luminosity should be proportional to the fourth power of terminal velocity. In other words: if galaxy A is sixteen times more luminous than galaxy B, its rotation curve will end up at a value that's twice as high.

In 1977 astronomers Brent Tully and Richard Fisher had already discovered a simple mathematical relationship between the luminosity and the rotational properties of spiral galaxies. From a dark matter point of view, the Tully-Fisher relation is somewhat surprising, since a galaxy's energy output is obviously dominated by stars, while its dynamics should be largely governed by the inventory of dark matter. Why would these two always conspire to yield the same relationship?

MOND provides a no-frills explanation, and while the Tully-Fisher relation hadn't been precisely calibrated in the early 1980s, subsequent observations have shown that it does indeed obey Milgrom's fourth-power prediction. Likewise, MOND naturally reproduces a similar relation for ellipticals (known as the Faber-Jackson relation).

But MOND is not some panacea. Even if we accept Milgrom's modified gravity, galaxies in clusters are moving too fast. Quite a bit of additional matter would still be needed, albeit not the huge quantities that dark matter theorists have proposed. And according to MOND, this matter would of course all be familiar matter. Initially some MOND scientists suggested that neutrinos might fit the bill, but other forms of matter would also do. X-ray astronomers had already discovered that galaxy clusters are filled with tenuous, hot gas, and that the mass of this intracluster gas is much larger than the combined mass of all the galaxies in the cluster. Who knows, there might be a comparable amount of dark, cold gas, invisible at any wavelength.

A more serious objection to MOND is that it's not a relativistic theory, at least not in its original incarnation. MOND was presented as an extension of Newtonian dynamics, not of general relativity. In other words, it had nothing to say about cosmic expansion, gravitational lensing (the bending of light in a gravitational field, the topic of the next chapter), black holes, and other phenomena that are so beautifully described by Einstein's general theory of relativity.

Only in 2004 did Milgrom's Israeli friend and colleague Jacob Bekenstein of the Hebrew University in Jerusalem write up a relativistic version of MOND, called TeVeS (for tensor / vector / scalar).[7] However, in its original formulation, TeVeS was frustratingly complex, it lacked the natural beauty of general relativity, and it has since fallen out of favor because it is inconsistent with recent observations of gravitational waves.[8]

So how seriously should we take MOND? Not at all, according to leading theoretical astrophysicists and particle physicists like David Spergel and Michael Turner, who have been attacking the idea for decades. Joel Primack once remarked, "If other cosmologists want to waste their time on MOND, that's great—it means less competition for me." And Jerry Ostriker says, "MOND gets everything wrong, except the rotation curves of galaxies, but that's just one piece of evidence. I never bothered to write a paper explaining why MOND cannot be right. It would be like explaining why humans can't fly."

The eighty or so attendees of the 2019 Bonn workshop clearly had a different opinion. The workshop, called "The Functioning of Galaxies: Challenges for Newtonian and Milgromian Dynamics," was mainly attended by adherents of MOND—a mixed bunch of predominantly male scientists from all over the world, who strongly believe that dark matter can't be the final answer. One of them, Constantinos Skordis of the Institute of Physics of the Czech Academy of Sciences, even presented a new relativistic theory of MOND.[9]

Milgrom has never been alone in his heroic and heretical fight against dark matter dogmatism. Almost from the very start, he has received strong support from others, including early converts Robert Sanders of the University of Groningen and Stacy McGaugh of Case Western Reserve University in Cleveland. "And things are getting better," Milgrom says. Yes, the older generation of astronomers has become very entrenched in its beliefs and convictions—as McGaugh once told me, "When I ask people what proof I could offer to convince them of MOND, they often say 'none.'" But the Bonn workshop attracted a lot of younger people, many of whom weren't even born when MOND was first proposed in 1983. These researchers are much more open to the unconventional idea. After all, since particle physics experiments have so far failed to detect dark matter, scientists are forced to think of alternative theories anyway. Suddenly, modifying gravity doesn't sound that crazy anymore.

McGaugh has become one of the most vocal proponents of modified Newtonian dynamics.[10] In 2020 he wrote a comprehensive review article for the open access journal *Galaxies,* listing an impressive number of cases in which MOND's predictions about the dynamics of rotating galaxies have subsequently been corroborated by observations.[11] In most cases, dark matter has so far failed to make similar successful predictions. As McGaugh concludes, "MOND gets all these predictions correct, in advance of their observation, because there is something to it."

MOND's most impressive achievement by far is its detailed prediction of the rotation curves of disk galaxies, purely on the basis of the observed distribution of baryonic matter (that is, stars and gas).

It's pretty incredible. You just give these guys any galaxy—big or small, compact or diffuse, extremely regular or highly chaotic—and using what they call the radial acceleration relation, they will compute its rotation curve for you. For hundreds of individual galaxies, these predictions have turned out to precisely match the observed rotation curves.

From a "Milgromian" viewpoint, the mathematically precise relation between rotation curves and the distribution of stars and gas is the most natural thing in the world. After all, according to MOND, galaxy dynamics are governed by baryonic matter only. But for dark matter theorists, that same relation is nothing less than a miracle. Sure, for each individual case you can assume a particular distribution of dark matter that produces the observed rotation curve, but it's a complete mystery why these rotation curves should be so exactly correlated to the distribution of baryonic matter. "If I want to predict velocities in a galaxy, I use MOND," says McGaugh. "That's the only thing that works. The fact that MOND works at all is a problem for dark matter. Why does this stupid theory get any prediction right?"

Before we go back into the main auditorium of the University of Bonn's Argelander Institute for Astronomy for the next talk, Milgrom has one more important thing to say. "Dark matter will never be falsified," he says. "If people don't find it, they can always claim that they haven't looked hard enough, and as long as you have enough free parameters, you can always assume that dark matter is distributed in precisely the way you need to explain the observations. In contrast, MOND could easily be proven wrong—for instance, if our interpretation of a certain galaxy's rotation curve would predict less baryonic matter than is actually observed. But this has never happened."

Or at least, in the few cases where other astronomers believe MOND has been ruled out (we'll get back to those in later chapters), Milgrom and his fellow heretics have always been able to come up with an explanation that saves modified gravity. Moreover, the concordance model of cosmology also has its problems (see chapter 22), but no one says that's reason enough to discard the theory altogether. "In the end, it's always a matter of which theory makes the most sense," says Milgrom.

In 2007, when I interviewed him for *Sky & Telescope,* the founding father of modified Newtonian dynamics expected the issue to be settled within twenty years or so.[12] Back then, Milgrom was sixty-one years old, and he told me he hoped to see an answer within his lifetime. I hope so, too—not just for him—but I'm not so sure. Even if dark matter doesn't exist, it has a powerful grip on the minds of most astrophysicists and cosmologists. As Robert Sanders wrote in his book *The Dark Matter Problem,* science is essentially a social activity, and if a whole community is misguided, it can be extremely hard to alter conventional wisdom.[13]

Some would say that MOND is a silly, contrived idea, like the luminiferous ether or the flat Earth. But it could also be a great new concept, like heliocentrism or continental drift. For now, we're all in the dark.

13

Behind the Lens

Not much more than twenty years ago, the road to the world's most powerful astronomical observatory was still an eighty-kilometer stretch of rocks, gravel, and potholes, leading south-southwest from Chile's Ruta Cinco through an eerie, Mars-like landscape. Our bus slowly made its way, shaking and rattling its passengers—a mix of astronomers, officials, and journalists.

It was March 5, 1999, and we were attending the inauguration of the European Southern Observatory's Very Large Telescope (VLT), a quartet of identical instruments perched on the 2,635-meter high summit of Cerro Paranal in Chile's bone-dry Atacama Desert. The four telescopes, each in its own thirty-meter tall cylindrical enclosure, can operate separately or join forces to create the sharpest possible views of the universe through interferometry, a technology inherited from radio astronomy.

At the time of the inauguration—by Chilean president Don Eduardo Frei Ruiz-Tagle, who flew in by helicopter—astronomers and engineers had carried out test observations with Unit Telescope 1. Now its first real science run was about to start. Meanwhile the second telescope had just seen first light, and UT3 was nearing completion. The enclosure of number four was still very much under construction.

It was a fun party, with great food and wine, but the preceding days had been even better thanks to their scientific rewards. Some 140 kilometers north of Paranal, at the Universidad Católica del Norte in the bustling harbor city of Antofagasta, dozens of researchers had gathered for the four-day symposium Science in the VLT Era and Beyond. Here, astronomers discussed the new facility's plentiful prospects and promises, and presented the very first results from the commissioning phase of UT1, whetting the appetite of astrophysicists and cosmologists.

One of the presentations, by Spanish astronomer Roser Pelló, was titled "Probing Distant Galaxies with Lensing Clusters." Pelló described spectroscopic observations of an extremely remote galaxy in the southern sky. As she explained, the redshift measurements—which put the object at a distance of some 11.5 billion light-years—were possible only because the galaxy's faint image is distorted and amplified by the gravity of a massive cluster of galaxies in the foreground, known as 1E 0657-558—a phenomenon known as gravitational lensing.

Little did I know that two decades later, this cluster, thanks to its gravitational lensing properties, would have made it into almost every astronomy textbook as convincing proof of the existence of mysterious dark matter. In fact, 1E 0657-558, better known as the Bullet Cluster, is generally described as the nail in the coffin of modified Newtonian dynamics—although not everyone agrees, as you might expect after reading the previous chapter.

The concept of gravitational lensing as a result of the curvature of spacetime has a long history. Albert Einstein first thought about gravity's ability to bend light paths in 1912, three years before he penned his general theory of relativity. His predictions were famously confirmed during the May 29, 1919, total eclipse of the Sun, when British astronomer Arthur Eddington established that stars close to the Sun's eclipsed disk were slightly shifted away from their expected positions, as if seen through a magnifying glass.

Much later, Einstein received a tip from an unlikely source: a Czech immigrant and former engineer named Rudi Mandl, who was earning his living as a dishwasher at a restaurant in Washington, DC. During a visit to Princeton, Mandl asked Einstein to figure out what the result of gravitational light bending would be if two stars were perfectly aligned with each other and the Earth. Light from the more distant star *A* emitted directly toward the Earth would of course never reach us—it would hit the far side of *B,* the star in the foreground. But light emitted from *A* traveling in a slightly different direction might pass close to *B,* be deflected by *B's* gravity, and eventually arrive on Earth, Mandl reasoned.

Intrigued, Einstein did some calculations and published the results in a brief 1936 note in *Science.*[1] "It follows from the law of deviation," Einstein wrote, "that an observer situated exactly on the extension of the central line *AB* will perceive, instead of a point-like star *A,* a luminius [*sic*] circle . . . around the center of *B.*" In other words: the background star would appear as a tiny ring of light encircling the foreground star.

Unfortunately, Einstein's calculations also showed that this "most curious effect" would never be actually seen: the apparent diameter of the ring of light is much too small. But just one year later, Fritz Zwicky argued that the phenomenon might be visible when looking at objects other than stars. Specifically, galaxies—"nebulae," as he continued to call them—offered a good chance to observe gravitational light-bending, with the strength and the geometry of the effect depending on the mass of the foreground object.

Zwicky described his idea, and coined the term "gravitational lens," in a letter to the editor in *Physical Review.*[2] He also realized the potential of gravitational lensing for the study of dark matter. Remember that his work on the Coma Cluster, described in chapter 3, had suggested that galaxies could be much more massive than you would guess on the basis of their visible contents. According to Zwicky, "Observations on the deflection of light around nebulae may provide the most direct determination of nebular masses and clear up the . . . discrepancy."

The mathematics suggested that a perfect ring of light—an Einstein ring, as it is called today—only shows up when the background light-emitting object and the foreground lensing object are precisely aligned point sources. In other cases—such as the less-than-perfect alignment of extended objects like galaxies—you might end up with multiple images of what is in fact one object in the sky, or with arc-like ring fragments. However, this all remained theoretical for decades. It wasn't until 1979, five years after Zwicky's death, that British radio astronomer Dennis Walsh and his colleagues hit upon the first gravitational lens: what appeared to be a twin quasar in the constellation Ursa Major.[3]

Quasars (quasi-stellar radio sources) are the luminous cores of remote galaxies. They are found all over the sky, but it is extremely unlikely to find two very similar ones so close together. Billions of light-years from Earth but separated by an apparent distance of less than 6 arc seconds, the twins were comparable to the headlights of a car some 300 kilometers away. Since such an arrangement is so unlikely, Walsh and his colleagues were not confident that they actually were looking at two individual quasars. However, in their discovery paper in *Nature,* they didn't jump to conclusions about what they were seeing.

Later observations confirmed their hunch that the two quasars are in fact one single object, apparently doubled as a result of gravitational lensing caused by a faint galaxy that lies between the quasar and the Earth. Long-exposure photographs made with large telescopes have since revealed the much fainter galaxy in question.

In the years since the discovery of the first gravitational lens, astronomers have found many more, including Einstein crosses (four images of a single background source), elongated light arcs (the greatly distorted images of distant galaxies), and even complete Einstein rings. Today such "strong" gravitational lenses are observed on a routine basis, in particular by the eagle-eyed Hubble Space Telescope. Most of them are found in rich clusters of galaxies, where the combined mass of the cluster does most of the light-bending, while individual galaxies are responsible for the peculiar details of the resulting scene.

But image multiplication, light arcs, and rings are just the conspicuous tip of the proverbial iceberg. Apart from strong gravitational lensing, there's also weak gravitational lensing, which induces lesser distortions in images of background galaxies. This can be the result of all sorts of intervening gravitating matter; tenuous gas, for example, can be found throughout intergalactic space, so spacetime is never completely "flat." The result, as astronomer James Gunn had already realized back in 1967, is that every distant galaxy's image is distorted to a certain degree.[4]

Through weak lensing, scientists can estimate the amount of gravitating mass—including dark matter—in a certain region of space. That's because the foreground mass slightly magnifies and stretches the images of faint, remote background galaxies, and the amount of distortion tells you how much mass does the gravitational lensing. This is not as simple as it sounds, though. Even in the absence of lensing, galaxies have elongated shapes, both because they're generally flattened and because we don't always see them face-on. So with just one galaxy, it's impossible to distinguish how much of its observed elongation is due to weak lensing. Instead, astronomers study as many background galaxy images as possible, looking for a tiny departure from the expected random distribution of galaxy elongations.

So here's the general idea: Observe hundreds (or thousands, or even millions) of faint background galaxies. Check for departures from random orientations. Use these departures to map the strength of the weak lensing effect that's responsible for the minute distortions. Then derive the corresponding mass distribution in the foreground. Presto: you've just arrived at a mass map of part of the universe. And since most of the universe's gravitating mass is dark matter, the map you've produced basically charts the dark matter along the line of sight—a feat first achieved (albeit with rather poor accuracy) by Anthony Tyson of AT&T Bell Laboratories and his colleagues in 1984.[5]

Maxim Markevitch of the Harvard-Smithsonian Center for Astrophysics hadn't thought too much about dark matter and mass distribution maps when he used the newly launched Chandra X-ray Observatory in October 2000 to observe the remote Bullet Cluster of galaxies, located in the southern constellation Carina, the Keel. But eventually, when combined with weak lensing research, this study would foster new dark matter knowledge.

Cluster 1E 0657-558, the Bullet Cluster, was known to be a bright X-ray source, indicating that it must contain massive amounts of extremely hot gas. Moreover, earlier observations had revealed that it consists of two separate subclusters, possibly in the process of merging. Reason enough for a detailed follow-up with NASA's new flagship X-ray satellite. Indeed, the Chandra image, published in 2002, exceeded all expectations. It clearly showed that the hot X-ray-emitting gas does not surround the largest galaxy concentrations, as is normally the case in galaxy clusters. Instead, the gas is concentrated between the two subclusters. Moreover, the image revealed what Markevitch and his colleagues called "a textbook example of a bow shock."[6] The bow shock was seen propagating in front of a bullet-like gas cloud—a feature that inspired the double cluster's popular nickname.

The interpretation was straightforward. Two smaller clusters must have collided almost head-on, with a mutual velocity exceeding 4,000 kilometers per second. The individual galaxies in the two clusters were not too much affected by the collision: the clusters were slowed by gravity, but the galaxies comprising them, generally separated by millions of light-years, easily passed each other, like bees in two sparsely populated swarms moving in different directions. However, the space between the galaxies in each cluster was filled with huge amounts of hot, tenuous intracluster gas, and these two gas clouds must have collided, as evidenced by the bullet-shaped bow shock. Due to the resulting ram pressure, the gas got swept out of the two clusters and piled up in the space between them.

So far, so good. But what about the invisible dark matter that is also supposed to be present in the two colliding clusters? Did the dark matter of the two clusters also collide, interacting like the baryonic gas particles? If so, you would expect that dark matter, too, got swept up and displaced to the region in between the two clusters. But what if dark matter particles are "collisionless"? What if they pay little attention to each other, just like the individual galaxies of the two clusters or the buzzing bees passing each other by? In that case—the case that most theorists expected—dark matter would have continued to accompany the galaxy concentrations, even as the two clouds of intracluster gas went their own ways.

Markevitch and his colleagues realized that the Bullet Cluster presented an ideal testbed to learn more about the properties of dark matter. "There is a clear offset between the centroid of the bullet subcluster's galaxies and its gas," they wrote. "If one measures the location of the subcluster's dark matter density peak (e.g., from weak lensing . . .), one may determine whether the dark matter is collisionless, as are the galaxies, or whether it experiences an analog of ram pressure, as does the gas."

Unfortunately, Markevitch didn't know too much about weak lensing. To many astronomers, the technique still seemed a bit like magic. But at a 2002 meeting on galaxy clusters in Taiwan, he ran into Douglas Clowe, a postdoctoral researcher at the University of Bonn.[7] Clowe was one of just a dozen or so weak lensing experts in the world at the time. Would he be interested in figuring out the mass distribution of the Bullet Cluster? Of course Clowe jumped on the opportunity.

For a convincing weak lensing analysis, you need to study the precise shapes of at least hundreds of faint background galaxies. That's only possible using a high-resolution photograph, obtained with a large optical telescope. Clowe figured it would be difficult and time-consuming to apply for observing time on an 8-meter-class telescope, so he decided to search astronomical archives for existing images of the cluster. That's how he came across the photo of the cluster shot during the commissioning phase of Unit Telescope 1 of the European Very Large Telescope. The very same photo I had been looking at during Roser Pelló's presentation at the VLT Opening Symposium in Antofagasta in March 1999.

Clowe's meticulous weak lensing analysis left little doubt about the nature of dark matter. The mass distribution map of 1E 0657-558 clearly shows two prominent peaks, coinciding with the two clusters. Evidently, the majority of the gravitating mass—the dark matter—still clumps around the cluster galaxies, indicating that the mysterious stuff is indeed collisionless. Great—dark matter appears to behave as most theorists expected of nonbaryonic particles like WIMPs.

But there was more. Around 2001, Clowe had attended a talk by Stacy McGaugh about modified Newtonian dynamics. While working on the weak lensing analysis, Clowe realized that the Bullet Cluster could hold the key to refuting MOND. If there's no mysterious dark matter in the universe at all, as MOND asserts, galaxies and intracluster gas is all there is. And since the total mass of the hot X-ray-emitting gas greatly exceeds the total mass of the visible galaxies, you would expect the weak lensing mass peak to coincide with the gas, not with the galaxies.

In an April 2004 paper in *The Astrophysical Journal*, Clowe, Markevitch, and Anthony Gonzalez of the University of Florida boldly concluded that the mass reconstruction of the Bullet Cluster provided direct evidence for the existence of dark matter.[8] Even if MOND were true, you would still need significant amounts of nonbaryonic dark matter to explain the observations. The authors concluded, "While these observations cannot disprove MOND . . . , they remove its primary motivation of avoiding the notion of dark matter."

A much more thorough analysis, published in 2006, arrived at the same general result.[9] This time Clowe and his colleagues used new cluster observations carried out in 2004 with the 6.5-meter Magellan telescope at Las Campanas Observatory in Chile and with the Advanced Camera for Surveys on board the Hubble Space Telescope. Because of the much better quality of the new data, the statistical significance of the result was now three times greater than before.

The Hubble observations and their implications drew much more attention than the original 2004 publication. An August 2006 NASA

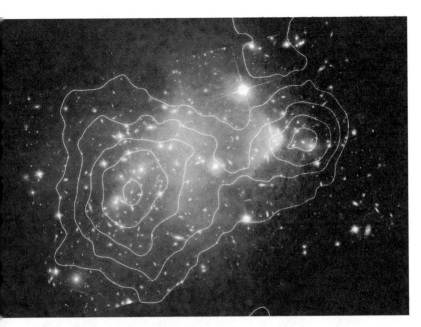

Hot, X-ray emitting intergalactic gas (diffuse glow) has piled up in the space between the two colliding subclusters of galaxies that make up the Bullet Cluster, while dark matter (contours, derived from weak lensing measurements) still clumps around the cluster galaxies.

press release quoted Clowe as saying, "These results are direct proof that dark matter exists."[10] And without mentioning MOND, the same press release stated that the weak lensing result "gives scientists more confidence that the Newtonian gravity familiar on Earth and in the solar system also works on the huge scales of galaxy clusters." Newspapers and popular science magazines ran a colorful combined image generated by the Chandra and Hubble telescopes, along with jubilant stories about the impressive result. For decades we had assumed that the universe is governed by dark matter; now, we had incontrovertible evidence.

Or did we? Did the gravitational lensing results on the Bullet Cluster kill MOND? Not at all, according to the adherents of modified gravity. Yes, MOND also requires some sort of invisible gravitating matter, but it could be in the form of neutrinos, or cold, compact objects that would be as collisionless as the galaxies in the colliding cluster. Moreover, some MOND-like theories offer a rather natural explanation for the Bullet Cluster's observed mass distribution, as Hongsheng Zhao of the University of St. Andrews, Scotland, told the participants of the 2019 Bonn workshop mentioned in the previous chapter.

Astrophysicist Robert Sanders of the University of Groningen credits a large part of the Bullet Cluster's proof-of-dark-matter hype to what he describes as NASA's formidable public-relations machinery. In his 2010 book *The Dark Matter Problem,* Sanders wrote, "It should be remembered that the implicit underlying assumption is that general relativity is the relevant theory of gravity on these scales . . . ; as with rotation curves the inferred existence of dark matter is not independent of the assumed law of gravity." In other words, dark matter is still a theoretical concept dreamed into existence in order that another theory—our ideas about the workings of gravity—can make sense. Findings made on the basis of gravitational-lensing analysis are consistent with the existence of dark matter, but they don't establish with certainty that the stuff must really be out there. "The proof of the existence of non-baryonic dark matter can only come with its direct detection," Sanders concluded.[11]

Which is not to say that gravitational lensing is unimportant in the study of dark matter. To the contrary, in a comprehensive 2010 review article in *Reports on Progress in Physics,* British astronomers Richard Massey, Thomas Kitching, and Johan Richard described gravitational lensing as "the most effective technique with which to investigate [dark matter]."[12] For instance, observations of strong gravitational lensing (multiple images and light arcs) can be compared with detailed predictions of the mass distribution within foreground galaxy clusters. Such checks can then be used to discriminate between different theoretical models—something that has been done successfully for six massive clusters in the Hubble Frontier Fields Program.[13]

Then there's galaxy-galaxy weak lensing, where tiny distortions in the shapes of background objects are caused not by the gravity of a whole cluster, as in the case of the Bullet Cluster, but by the light path–bending power of a single, massive foreground galaxy. This type of lensing tells astronomers about the dark matter halos thought to surround visible galaxies. So far, observations seem to support the existence of these halos.

Yet another form of weak lensing, and one that may enable astronomers to map the distribution of dark matter throughout the observable universe, is cosmic shear. Cosmic shear is caused when background light passes through the uneven distribution of matter across space, as Jim Gunn predicted in 1967. The universe's large-scale structure—the clusters, superclusters, and great walls of galaxies, as well as the vast, mostly empty voids—slightly influences the trajectory of every single light ray passing through the depths of space. It's like an uneven floor that causes a marble to roll along a wiggling path instead of a straight line. By the time a distant galaxy's light arrives at our telescopes, even if there's no galaxy or cluster right in front of it, its shape is nonetheless slightly distorted. That distortion provides you with information about the distribution of both visible and dark matter along the light ray's path.

To observe cosmic shear, you need to statistically analyze the shapes of millions of remote galaxies in wide swaths of sky. This only became possible with the advent of a new digital image-sensor technology called a charge-coupled device. Cosmic shear was first observed by no fewer than four separate teams of astronomers, who all published their results in May 2000.[14] By measuring cosmic shear in galaxies at various distances in the universe, astronomers can also apply a technique known as cosmic tomography, yielding something like a 3D MRI scan of all the mass in the universe. Cosmic tomography makes it possible to examine the large-scale structure of both baryonic and nonbaryonic matter at different distances, corresponding to different look-back times, as we'll see in chapter 22.

We live in a universe in which light is deflected by gravity, photons surf the waves of spacetime, galaxies get distorted, and nothing is what it seems. Yet the giant eyes that astronomers have constructed to spy on the universe are starting to show what is really going on. During my most recent visit to the Very Large Telescope, in September 2018, the observatory was still in full swing, almost twenty years after its inauguration. So too was the study of gravitational lensing, which has become an important aspect of cosmological observation. We haven't yet solved all of dark matter's riddles, but we're making progress, and when we reach the limits of the current generation of telescopes, more powerful instruments will take the baton, eventually revealing the large-scale distribution of dark matter in the process.

Looking east from the VLT observing platform, across 20 kilometers of arid desert, I can easily make out the flattened summit of Cerro Armazones, where the European Southern Observatory is constructing its next flagship facility: the Extremely Large Telescope.[15] For decades to come, this monstrous and versatile telescope, with a primary mirror almost 40 meters wide, will be the best available ground-based tool to study gravitational lenses in detail.

I can hardly wait to attend its inauguration.

14

MACHO Culture

OK, so most of the gravitating matter in the universe is dark. As we've seen, the evidence is overwhelming. Moreover, the very composition of the visible universe (in particular, its deuterium content) tells you that there's way too little baryonic matter—atoms—to explain the mystery. Case closed, it would seem.

But wait a moment. The fact that the dark matter can't all be baryonic doesn't necessarily mean that all of the dark matter has to be nonbaryonic. At smaller scales, more familiar forms of invisible matter cannot be ruled out completely. For instance, extremely dim dwarf stars, or even rogue planets, could populate the extended halos of galaxies like our own Milky Way, just as Jerry Ostriker, Jim Peebles, and Amos Yahil had suggested in 1974.

At least, that was the view of a substantial number of astrophysicists in the late 1980s. Maybe they were just sensibly conservative, careful not to throw the baby out with the bathwater. In any case, they believed that maybe the flat rotation curves of galaxies could be produced not by hypothetical, wimpy, and weakly interacting elementary particles but rather by much larger and more massive objects of the astrophysical kind. Not by WIMPs but by MACHOs—massive compact halo objects.

MACHOs could be red dwarfs—puny stars, much smaller, cooler, and fainter than our own Sun. Or they could be brown dwarfs: even smaller balls of gas that are neither massive nor hot enough to fuse hydrogen in their cores. Old white dwarfs, the compact leftovers of Sun-like stars that slowly cool down and fade with age, could fit the bill. So could super-dense neutron stars or small black holes, the mortal remains of very massive stars. Small primordial black holes, left over from the big bang, might also be among the MACHOs. And yes, countless orphan Jupiter-like planets, much too small and dark for any form of direct detection, might swarm around in the halos of galaxies.

But if MACHOs are too faint to see—after all, we're talking dark matter here—how could astronomers possibly prove their existence? The surprising answer is: through gravitational lensing. Remember how Einstein had calculated how the light of a distant star can be bent and amplified by the gravity of a precisely aligned foreground star? The same effect could be achieved due to an invisible foreground MACHO. In other words, if a MACHO in the halo of our Milky Way passed in front of a much more distant star (in another galaxy), the MACHO would temporarily change the appearance of the background star in a very specific way.

Yes, Einstein wrote that "there is no hope of observing this phenomenon directly," but that was in 1936. He was right about the ring-of-light-effect: in the case of individual stars, the resulting ring is way too small to observe, even with the largest telescopes. But the foreground object—the gravitational lens—also amplifies the light of the background star, and in the case of a near-perfect alignment, the amplification can be substantial, as Norwegian astrophysicist Sjur Refsdal noted in a 1964 paper in *Monthly Notices of the Royal Astronomical Society*.[1] Refsdal showed that the passage of a foreground star in front of a background one, as seen from our terrestrial vantage point, isn't that rare a phenomenon at all. According to Refsdal, they occur "rather frequently. The problem is to find where and when the passages take place."

This was fifteen years before the first gravitational lens was discovered—the "twin quasar" described in the previous chapter. And Refsdal was writing long before anyone had ever worried about dark matter halos. But in 1981, PhD student Maria Petrou realized that invisible bodies in the halo of the Milky Way might give themselves away through gravitational lensing of extragalactic background stars. Unfortunately, her supervisor prevented her from publishing her results.[2]

Born in Greece, Petrou did her undergraduate education at the University of Thessaloniki, then studied mathematics and astronomy at Cambridge. Her thesis, "Dynamical Models of Spheroidal Systems," dealt with a number of theoretical topics, including "the gravitational lens effect due to the halo objects of our galaxy." She investigated "what one might expect to see if our Galaxy has a halo made from 'Jupiters,' white dwarfs or black holes." Petrou concluded "that if our Galaxy possesses a halo of dark compact objects . . . , we might be able to see transient amplification of extragalactic stars, the duration of which depends upon the kind of halo objects acting as lenses."

Petrou's ideas were prescient, but they didn't reach the wider astronomical community at the time. Although most of the chapters of her thesis were published in peer-reviewed scientific journals, her thesis advisor, the famous Cambridge astronomer Donald Lynden-Bell, thought the chapter on halo objects was too speculative and didn't let her submit it for publication.

Five years later, in May 1986, Polish-born Princeton astronomer Bohdan Paczyński published a landmark paper, "Gravitational Microlensing by the Galactic Halo," in *The Astrophysical Journal*.[3]

According to Paczyński, if the Milky Way's dark matter halo consists of objects with masses larger than about half the mass of the Moon, any star in a nearby galaxy has a one in a million chance of being "microlensed" at any time. What this means is that, as the halo object passes in front of the background star, the star's light would be amplified for a couple of days, weeks, or months, depending on the lens mass—just as Petrou had shown. The star's brightness would rise to a maximum value before falling off again in a perfectly symmetrical way.

Of course, you never know in advance which background star will be lensed, so the only way to find a relevant instance of lensing would be to continuously monitor many millions of stars for an extended period of time—say, two years or so. "The data-processing aspect of the suggested observational program seems formidable," Paczyński wrote. Indeed, working from photographic plates would be very time-consuming; in the mid-1980s, electronic detectors were small and unable to image enough stars in one exposure, and astronomers had relatively little experience with automated data processing. Like Petrou, Paczyński was ahead of his time. In fact, the referee of his paper initially advised against publication because his ideas lacked any practical applicability.

Three years later, things looked much more promising. Charles Alcock, an astronomer at the Lawrence Livermore National Laboratory in California, was working on a project to automatically monitor large numbers of stars, together with his colleague Tim Axelrod and postdoc Hye-Sook Park.[4] The plan was to search for distant solar system objects like frozen bodies in the Kuiper Belt and cometary nuclei in the Oort Cloud, which would betray themselves by briefly blotting out the light of a distant star when passing in front of it. A rare and utterly fleeting event, but by keeping track of the brightnesses of thousands of stars using electronic detectors and special-purpose software, you might just be lucky enough to catch one.

In 1989 Dave Bennett, a postdoc at Princeton, urged Alcock to have another look at Paczyński's 1986 paper. If you're able to see brief stellar winks due to passing Kuiper Belt Objects and comets, Bennett suggested, you may also be able to detect much slower brightness variations due to the microlensing effects of passing dark bodies in the galactic halo. Alcock immediately got excited. It didn't take all that long to work out most of the details, and on October 31, during a seminar at the Center for Particle Astrophysics (CfPA) at the University of California, Berkeley, he presented the new plan for a microlensing survey that might solve at least part of the dark matter mystery—by showing that some of it, at least, was not mysterious at all but rather was ordinary baryonic matter that had simply eluded detection.

The collaboration with the Berkeley scientists, just down the road from Alcock's perch at Livermore, made a lot of sense. The CfPA was developing detectors to look for particle dark matter—WIMPs, basically—and the center's newly appointed director, French physicist Bernard Sadoulet, was eager to invest money and manpower in the microlensing project. If it was successful, Berkeley would share credit for a major discovery; if not, the absence of baryonic dark matter would strengthen the case for the center's WIMP search. Moreover, CfPA physicist Chris Stubbs, a tech wizard, was experimenting with connecting individual charge-coupled device (CCD) detectors into large-format mosaics that would yield a much wider field of view, capturing more stars at once. Such a device would be extremely useful for Alcock's research.

Now the challenge was to find a sufficiently large telescope that also could be used as a dedicated survey instrument for years in succession. The telescope had to be in the southern hemisphere, since the best possible survey area would be the Large Magellanic Cloud, a small companion galaxy to our own Milky Way that is invisible from most of the northern hemisphere. At a distance of just 167,000 light-years, the Large Magellanic Cloud is so close that observers on Earth can see its individual constituent stars, and dense enough to provide many potential microlensing sources. Its smaller neighbor, the Small Magellanic Cloud, was a valuable secondary target.

As luck would have it, such a telescope was available. Old and broken down, with a 1.27-meter mirror, the telescope was sitting unused in a dome at the Mount Stromlo Observatory, near Canberra, Australia, as Bennett learned when he approached Mount Stromlo astronomer Ken Freeman for help finding a suitable survey instrument. Known as the Great Melbourne Telescope, it dated back to 1868, when it was the largest fully steerable telescope in the world. It was moved to Mount Stromlo in 1947.

Freeman had been interested in dark matter since his work on galactic rotation curves in 1970 (see chapter 8), and during a 1985 stay at the Institute for Advanced Study in Princeton, he had learned about Paczyński's theoretical work on microlensing. Wouldn't it be wonderful if the Great Melbourne Telescope could be restored to its former glory and play the lead role in this exciting project? Together with his colleague Peter Quinn, Freeman talked observatory director Alex Rodgers into the plan, and in 1990 funds were being freed up to revitalize the telescope.

Thus, the Livermore-Berkeley-Mount Stromlo MACHO collaboration was born. The catchy acronym was thought up in 1991 by team member Kim Griest. Who needed WIMPs anyway?

But Alcock and his collaborators were not alone. Among the attendees of the Halloween 1989 CfPA seminar was James Rich of the Centre d'Études de Saclay in France. Back in Paris, Rich discussed Alcock's plans with his colleagues Michel Spiro and Éric Aubourg, who thrilled to the idea of finding dark halo objects through microlensing. Spiro, Aubourg, Rich, and others started their own project right away. It became known as EROS: Expérience pour la

The Great Melbourne Telescope during construction in 1869. In the 1990s, the telescope was used to search for MACHOs—massive compact halo objects.

Recherche d'Objets Sombres (Experiment for the Study of Dark Objects).[5]

Most of the Saclay scientists had a background in physics, and the cultural difference with astronomy was clear from the start. The EROS group wasn't intimidated at all by the prospect of weeding reams of data in search of extremely rare events—that was exactly what they were used to doing when analyzing the measurements of the Large Electron-Positron Collider, which had just come online at CERN. Saclay's computer nerds hammered out special-purpose software for automated data analysis, and the engineers at the particle physics department started work on the design and construction of a large electronic camera.

To further increase the chance of being first, the EROS team decided not to wait until their digital camera was ready. Instead, they started their microlensing program by taking old-fashioned photographs of the Large Magellanic Cloud—glass plates (really) that were digitized for subsequent computer processing. Through the connections of Alfred Vidal-Madjar of the institute's astrophysics department, the team got in touch with the European Southern Observatory (ESO) in hopes of using one of its telescopes to make the plates. The ESO operated no fewer than fourteen telescopes at their La Silla Observatory in Chile, including a 1-meter Schmidt telescope—the instrument of choice for taking wide-field images of the night sky. Already in 1990, the first ESO-Schmidt photographic plates of the Large Magellanic Cloud, showing around eight million faint stars, were being digitized using the Paris Observatory's novel MAMA densitometer, a plate-measuring machine that basically yields the visual brightness of each and every star in the image (MAMA stands for Machine Automatique à Mesures pour l'Astronomie). By comparing data from different nights during the campaign, the software should be able to find the occasional star that varies in brightness in a telltale microlensing way.

By late 1991, the new EROS 1 digital camera started operations on a telescope piggybacked on ESO's 40-centimeter GPO refractor (Grand Prisme Objectif), also at La Silla. With its 3.7 million pixels, the EROS 1 was the largest digital camera ever constructed at the time. Still, the field of view was smaller than the Schmidt camera's, so the electronic images contained fewer stars—about a hundred thousand as opposed to 8 million. But the digital camera's exposure times were much shorter, and, potentially, it would be able to detect shorter-duration events produced by lower-mass MACHOs.

Meanwhile, Alcock's MACHO collaboration was constructing its own digital camera. Given the French competition, time was of the essence—the discovery of a large population of dark objects in the Milky Way halo had Nobel Prize potential. Stubbs succeeded in hooking up four large CCDs into a 16.8-megapixel mosaic—another world record. However, partly because of the time-consuming work on the Great Melbourne Telescope, the first Magellanic Cloud images weren't obtained until July 1992. From that moment, the race was on.

(A third microlensing program, the Optical Gravitational-Lens Experiment, began in 1992 and is still running.[6] However, OGLE, which was initiated by Paczyński, focuses on objects in the central regions of the Milky Way galaxy and isn't specifically looking for dark matter, so I don't discuss it in detail here.)

Obviously, if you check millions of stars for brightness variations, you will find countless cases where microlensing plays no role at all. Many stars are naturally variable, for instance because they show regular pulsations. In other cases, brightness variations are due to the fact that the star is actually a binary—two stars orbiting the same center of mass, each eclipsing the other at regular intervals. Such periodic variables can safely be discarded, since microlensing produces one-off events. But there are isolated cases of rising and falling brightness, and you need to be sure that these, too, are not caused by some odd behavior of the star itself.

Luckily there are ways to make the distinction between microlensing and other sources of nonperiodic variation. One key characteristic of a microlensing event is the perfectly symmetrical shape of its light curve—the graph showing how brightness varies with time. If a star brightens at one rate and then dims at another, microlensing cannot be the cause. And there's another important clue. An intrinsically variable star usually changes color, however subtly, because its surface temperature rises and falls. The result is that the star's behavior as seen through a red filter is slightly different from what is seen through a blue filter. In contrast, microlensing events are expected to be "achromatic": red light is amplified in exactly the same way as blue light. Each of the competing groups was sure to check for possible color effects: using both the Schmidt photographic plates and the digital camera, EROS obtained successive exposures through red and blue filters, while the MACHO collaboration split the incoming light into two wavelength bands and used two identical cameras to observe the star field in both colors at the same time.

In the summer of 1993, it became clear that Alcock's MACHO collaboration had bagged its first event. Over the course of about a month, an utterly inconspicuous Large Magellanic Cloud star—one among many millions of others—had slowly brightened and faded again in a neat symmetrical fashion. The star reached peak brightness, seven times greater than normal, around March 11. From the duration of the event, the mass of the microlensing culprit was estimated at about one-eighth that of the Sun. After analyzing some 12,000 images, everything suggested that the first bona fide MACHO had revealed itself—a find that would surely warrant a *Nature* publication.

In September, while preparing the discovery paper, Alcock learned that the EROS collaboration was also about to publish its first results, based on an analysis of over 300 Schmidt plates and more than 8,000 CCD images. EROS had found two microlensing events, peaking around December 29, 1990, and February 1, 1992, respectively. Now the heat was really on. Alcock's team finalized their paper in under two days and submitted it to *Nature* on September 22, the same day the EROS paper, with Aubourg as its lead author, landed on the editor's desk. Both papers were accepted within a week and published back to back in the October 14 issue—unusually fast.[7]

"In Search of the Halo Grail," read the title of an accompanying commentary by University of Washington astronomer Craig Hogan.[8] Hogan hailed microlensing as "a powerful new technique to probe the composition of dark matter. . . . The microlensing programs may at last reveal the form and hiding place of most of the baryons in the Universe." However, he added, "with so few possible detections, the case for microlensing is not yet conclusive."

Indeed, within a couple of years, Aubourg and his coauthors had to retract each of their claims, much to their disappointment. Follow-up observations revealed that both of their suspect stars were weird variables after all, with long quiescent periods and occasional brightness changes that had all the symmetric and achromatic characteristics of microlensing events.

The EROS 1 camera operated until 1995. In June of the following year, the much larger EROS 2 camera, sporting 32 million pixels (again, the world's largest at the time), saw first light at La Silla—not in the old GPO dome but on the 1-meter Marly (Marseille-Lyon) telescope that had been relocated to Chile from the Haute Provence Observatory in Southern France. This time the team used a dichroic beam splitter to observe in two colors at the same time. But, despite EROS 2's higher sensitivity and larger field of view, and despite the better overall data quality it yielded, the camera failed to detect any new convincing microlensing events during its six and a half years of operation.

The MACHO collaboration fared somewhat better. Their first detection stood the test of time, and over the years, they found a number of other candidates. However, the interpretation wasn't always straightforward. Yes, these events may have been caused by brown dwarfs in the galactic halo, but they could also be due to low-mass stars in the outer regions of the Magellanic Cloud itself, in which case the results had nothing to say about Milky Way MACHOs.

Eventually, the enthusiasm surrounding compact halo objects dwindled. If the massive halo of the Milky Way galaxy were made of free-floating Jupiters, failed stars, or dark stellar remnants, the Large Magellanic Cloud should turn up something like ten microlensing events per year. Instead, only a handful were found over a period of almost a decade.

In May 1998, the two competing teams joined forces to publish a combined intermediate analysis of their results. Their *Astrophysical Journal* paper, "EROS and MACHO Combined Limits on Planetary-Mass Dark Matter in the Galactic Halo," concluded that there just aren't enough dark, compact objects to explain the flat rotation curves of galaxies.[9] In our own Milky Way, the teams determined, at most 25 percent of the derived halo mass can be accounted for by MACHO-like objects. Later, when more data became available, that limit was lowered even further.

The MACHO collaboration closed shop in late December 1999, ending what Alcock describes as "the highlight of my professional career." (He is now the director of the Harvard-Smithsonian Center for Astrophysics.) The team's final paper, wrapping up more than five years of measurements on almost 12 million stars, was published in October 2000 in *The Astrophysical Journal*.[10] Less than two and a half years later, on January 19, 2003, a devastating bushfire at Mount Stromlo destroyed the observatory, including the Great Melbourne Telescope.

EROS continued until February 2003 and presented an overview of their results in summer 2007 in the European journal *Astronomy & Astrophysics*.[11] Two years later, the Marly telescope was decommissioned and then shipped overseas. It is now the main instrument at a small observatory on Mount Djaogari in Burkina Faso, some 250 kilometers northeast of the capital Ouagadougou.

MACHO hunting has ceased. The WIMPs have won.

15

The Runaway Universe

On January 8, 1998, scientists announced that the universe will never stop expanding, but for some silly reason I missed the press conference. Yes, I was one of the journalists attending the 191st meeting of the American Astronomical Society in Washington, DC. But it was my first AAS, and I had a hard time figuring out what was happening at what time and in which of the many rooms at the Washington Hilton Convention Center. So when Saul Perlmutter and Peter Garnavich presented their exciting results on the never-ending growth of the cosmos, I was probably listening to some dull talk elsewhere in the building.

Ever since the big bang, empty space has been expanding. But scientists have not always been sure if this expansion will continue indefinitely. That's because cosmic expansion is slowed down by the collective gravity of all the matter in the universe. For decades, astronomers wondered whether there is enough gravitating matter—luminous and dark—to not just slow the expansion but bring it to a halt and ultimately reverse it. The result would be contraction, leading to what has been called a big crunch. Just how much matter is there, then? As we have seen in previous chapters, it's not easy to "weigh" the universe and find out. So two teams of researchers, Perlmutter's and Garnavich's, independently came up with another way to assess the matter content, and therefore the future, of the universe: they measured the expansion's history, by looking at distant supernova explosions.

The supernova message was loud and clear: there's not enough slowing down going on to ever stop cosmic expansion. Apparently the universe's biography is a never-ending story. As Perlmutter said at the AAS press briefing, "For the first time, we're going to actually have data, so that you will go to an experimentalist to find out what the cosmology of the universe is, not to a philosopher." At least, that's what I read from other reporters. The next day, the *New York Times* had the news on its front page. "New Data Suggest Universe Will Expand Forever," the headline read.

But there was more. The supernova measurements indicated not only that the expansion of the universe was endless but that it wasn't even slowing down. The rate of expansion was in fact accelerating. That result wasn't announced at the press conference. The discovery was so surprising, so weird, and so far-reaching that it took another six-and-a-half weeks before one of the competing research groups felt confident enough to announce it in public.

We live in an accelerating cosmos, a runaway universe. Empty space is being pushed apart by some uncanny force that cosmologists have christened dark energy, for lack of any better name. As if dark matter wasn't mystery enough—curiouser and curiouser, to quote Lewis Carroll's Alice. In December 1998, *Science* hailed the discovery of the universe's accelerating expansion as the scientific breakthrough of the year; in 2011, three key scientists behind the revolutionary find, including Perlmutter, shared the Nobel Prize in Physics. And although the true nature of dark energy still eludes astronomers and physicists alike, the runaway universe is here to stay. Forever and ever and ever.

To refresh your memory: cosmic expansion—the first indication that the universe must have had a beginning—was discovered in the 1920s. As mentioned in chapter 3, Vesto Slipher was the first to note that the light of most "spiral nebulae" is redshifted, indicating that they are receding from us at surprisingly high velocities. Georges Lemaître and Edwin Hubble later realized that recession velocities are larger for galaxies that are farther away, which is exactly what you'd expect if the universe as a whole is expanding—one of the possible outcomes of the equations of Albert Einstein's general theory of relativity. Before long, astronomers arrived at the big bang theory for the origin and early evolution of the universe.

The most instructive way to quantify the expansion of the universe is through its relative growth rate. It's a bit like monetary inflation. The inflation rate cannot be expressed as an absolute dollar amount; that would only work for a particular sum of money. Rather, inflation must always be given as a percentage. The same is true for cosmic expansion: it cannot be expressed in, say, kilometers per second or miles per hour, unless we are talking about the recession velocity *right now* of an object at a particular distance. It's much more useful to express the rate of comic expansion as a growth percentage per unit of time.

It turns out that cosmic distances do not grow very fast. In fact, they increase by some 0.01 percent in 1.4 million years. In other words, if the current distance to a remote galaxy is 100 million light-years, the distance will increase by one light-year every 140 years or so. A recession velocity of one light-year per 140 years corresponds to about 2,150 kilometers per second (4.8 million miles per hour). But this recession velocity only applies to the galaxy in question, and to other objects at a similar distance of 100 million light-years. A galaxy at a distance of 200 million light-years appears to recede twice as fast, at around 4,300 kilometers per second. The recession velocity grows by 21.5 kilometers per second for every additional million light-years.

This proportionality constant—21.5 kilometers per second per million light-years—is one way to quantify the expansion rate of the universe. Astronomers don't usually express cosmic distances in light-years, though. Instead, they use the parsec, where 1 parsec equals 3.26 light-years. So an astronomer will tell you that the universe is expanding at a rate of 70 kilometers per second per million parsec (70 km / s / Mpc), which is known as the Hubble constant. Using the Hubble constant, it's easy to convert recession velocities, as determined from galactic redshifts, into distances.

There's one catch: the cosmic expansion rate isn't truly constant, and neither is the Hubble constant. (For that reason, many astronomers prefer to call it the Hubble parameter.) The expansion of the universe is slowed down by the combined gravity of baryonic and nonbaryonic matter. This is a direct result of general relativity, which says that the overall behavior of spacetime is governed by the matter and energy it contains. As a result, you expect the expansion rate to decrease over time.

Now it becomes clear how the universe's fate is determined by its average density. In a high-density universe, gravity will eventually bring expansion to a halt. The universe then starts to contract again, heading for a big crunch. This model is called a closed universe, since spacetime closes in on itself. It's also referred to as a positively curved universe because its overall four-dimensional curvature is geometrically comparable to the three-dimensional curvature of a sphere.

In a low-density universe, the expansion slows down over time but never comes to a complete stop. In the distant future, matter is so diluted that gravity hardly plays a decelerating role anymore, and the universe keeps on expanding forever at a constant rate. This is called an open or negatively curved universe, with spacetime shaped like an infinite four-dimensional equivalent of a Pringles chip: curved in every possible direction, but never closing in on itself.

In between these two possibilities is the equilibrium case of a flat universe—a world model, as some cosmologists used to call it—without an overall curvature. In a flat universe, the density is *just* high enough to keep slowing down cosmic expansion forever but not high enough to cause reversal and contraction. This is called the critical density, a term we encountered in chapter 11. At present, the critical density equals approximately 10^{-29} grams per cubic centimeter.

After the introduction of the big bang theory, it seemed that two numbers would be enough to know the fate of the universe: the Hubble parameter—a measure of the current expansion rate—and the deceleration parameter, which tells you how quickly cosmic expansion is slowing down. In a famous 1970 *Physics Today* paper, "Cosmology: A Search for Two Numbers," Allan Sandage of the Mount Wilson and Palomar Observatories, wrote, "If work now in progress is successful, better values for [the Hubble and deceleration parameters] should be found, and the 30-year dream of choosing between world models on the basis of kinematics alone might possibly be realized."[1]

Back then, Sandage probably didn't expect it would be another three decades before his cosmology dream came true. That finally happened in May 2001, when researchers published results from a Hubble Space Telescope program that precisely measured the Hubble parameter. As we will see in chapter 22, though, astronomers and cosmologists are still debating its true value.[2] As for the deceleration parameter, well, that was the topic of the January 1998 AAS press conference I didn't attend. Twenty-eight years after Sandage's "Two Numbers" paper, scientists were convinced that the universe will never stop expanding.

Which is not to say that they didn't have their opinions and preferences in the intervening years. As mentioned in chapter 11, a flat, critical-density universe was an "aesthetically pleasing idea" to many astronomers, and for good reason. Observations of the distant universe had already suggested that any overall curvature—whether positive or negative—had to be relatively small, otherwise it would show up in galaxy counts. In a flat, Euclidean geometry, the number of galaxies in a certain area of sky grows with the square of the distance. As a result, remote, faint galaxies are much more numerous than nearby brighter ones. Although it's a subtle effect, a strongly curved universe would show measurable deviations from this inverse-square law.

So if our universe does have an overall curvature, it can only be barely open or else barely closed, otherwise we would have noticed it long ago. And that would be weird, to say the least. The big bang theory doesn't prescribe any particular curvature or geometry, so why would the universe be extremely close to flat, but not *precisely* flat? It seems much more likely that, for some reason, the curvature of the universe is exactly zero.

Thanks to the pioneering work of theoretical physicist Alan Guth, cosmologists believe they know what that reason is: inflation. According to Guth's inflationary hypothesis—which was conceived in late 1979, published in 1981, and subsequently amended and improved by Russian-American physicist Andrei Linde—the newborn universe went through a rapid phase of exponential growth in the very first 10^{-35} seconds of its existence, doubling in size about a hundred times in succession.[3] Such an incredible size boost would give the present observable universe a curvature that's indistinguishable from zero, no matter how strongly curved the geometry might have been at the very beginning. That's because the overall curvature of the exponentially expanding universe rapidly decreases with size, just like the curvature of the Earth is less conspicuous than the curvature of a marble.

You could write a whole book about inflation; in fact, Guth did just that, as did a number of other authors. But while the hypothesis solves a number of nagging cosmological problems, it's still rather speculative, and the technicalities don't bear that much on our topic of dark matter, so I won't go into great detail here.[4] Suffice it to say that inflation provided cosmologists with ample reason to assume that our universe is perfectly flat, which can mean only one thing: it must have the critical density. And since big bang nucleosynthesis tells us that baryonic matter can account for only 5 percent of the critical density, inflation seems to suggest that there must be really huge amounts of nonbaryonic dark matter out there.

Saul Perlmutter, a physicist at the Lawrence Berkeley National Laboratory, decided to find out just how much. Not by hunting for dark matter, as others were doing, but by actually measuring the deceleration of the universe's expansion. The faster the deceleration, the more matter—visible and dark—our universe must contain. (An empty universe would of course not show any deceleration at all.) To figure out the deceleration rate, Perlmutter searched for supernovae in distant galaxies, a program initiated by his Berkeley colleague Carl Pennypacker. By precisely comparing the apparent brightnesses of supernovae with their redshifts, it would be possible to gauge how rapidly the expansion of the universe is slowing down, a method Sandage and Gustav Tammann had already suggested in 1979.

Here's how it works. The redshift tells you the extent to which the wavelength of light from the supernova has been stretched during the trip through expanding space, which may have taken hundreds of millions or even billions of years. A precise redshift measurement could, for instance, reveal that the supernova exploded at a time when cosmic distances were 30 percent smaller than they are now. If the universe has always been expanding at the same rate—that is, if the Hubble parameter were truly a constant—this would immediately tell you the corresponding light travel time (remember, 0.01 percent growth in 1.4 million years).

However, in a decelerating universe, the expansion rate must have been larger in the distant past than it is now. That means that it took less time for the universe to grow from its former to its present size than would a "coasting" universe, with a constant expansion rate. In other words, the supernova's light travel time is smaller, corresponding to a smaller distance, so the explosion must appear brighter than you would naively expect from its redshift. Instead of a linear relation between redshift and perceived brightness, you would notice a deviation from strict linearity for really distant supernovae, and the brighter they appear, the stronger the deceleration of cosmic expansion must be, hinting at a higher-density universe.

This trick only works if your supernovae all have the same luminosity. That's why Perlmutter, Pennypacker, and their colleagues focused on one easily recognizable type of supernova, called Ia. Type Ia supernovae occur as white dwarf stars blow themselves into smithereens, for instance as a result of mass transfer from a close companion. When a white dwarf grows over 40 percent more massive than our Sun, pressure and temperature in its core build up high enough for carbon to ignite, and the star ends its life in a catastrophic thermonuclear detonation. Bad news for the star, but good news for cosmologists: since all exploding white dwarfs have more or less the same mass (1.4 times the mass of the Sun), all type Ia supernovae are expected to have more or less the same true luminosity. That makes it possible to really check how much brighter they appear than you would expect from their redshifts.

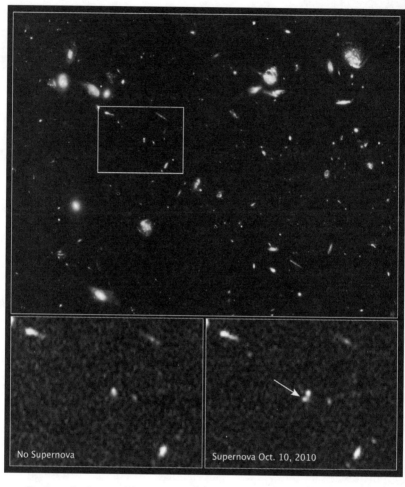

No Supernova

Supernova Oct. 10, 2010

By comparing images of the same part of the sky at different times, the Hubble Space Telescope discovered some of the most distant supernova explosions ever observed.

Over the years, using various telescopes and digital cameras, Perlmutter's Supernova Cosmology Project succeeded in measuring first one, then a dozen, and eventually over forty distant type Ia supernovae. Quite a feat, since supernova explosions are pretty rare. You don't know in advance when and where a new one will pop up. However, by observing tens of thousands of remote galaxies at once, chances were high that you would catch one or two exploding stars. The researchers would image their galaxies and then do so again a couple weeks later. Dedicated software would tease out the tiny pinpricks of light that showed up on the second image but not on the first—telltale traces of supernovae. Follow-up observations with other telescopes could then study the supernovae in more detail and determine their redshifts, an approach pioneered by Danish astronomers at the European Southern Observatory in Chile.

In the first half of the 1990s, it became clear that the Supernova Cosmology Project was starting to obtain interesting results. Other scientists began to take note. In 1994 Harvard astronomer Brian Schmidt (who moved to Australia in 1995) and Nick Suntzeff of the Cerro Tololo Inter-American Observatory in Chile started their own program, in the hope of beating the Berkeley physicists in their quest for the deceleration parameter. Before long, Schmidt and Suntzeff's High-z Supernova Search Team (the letter z is used to denote redshift) was also hunting for remote supernovae, using an approach similar to the Berkeley group's. Within a few years, the two teams were competing for precious observing time at the 3.6-meter Blanco Telescope at Cerro Tololo and the Hubble Space Telescope.

While Perlmutter's background was in physics, Schmidt and his collaborators, including his Harvard PhD supervisor Robert Kirshner, were astrophysicists and supernova experts. They were relatively late in the game of high-redshift supernova hunting, but they had much more experience in carrying out astronomical observations and in handling the nasty properties of type Ia supernovae, mainly by studying stellar explosions closer to home. In particular, it had become clear that Ia's did not always explode with exactly the same energy. And if you didn't know a supernova's true luminosity, it would be hard—if not impossible—to draw conclusions about its distance on the basis of its apparent brightness.

Most of these problems were eventually tackled, largely thanks to the work of Mark Phillips at Cerro Tololo and Adam Riess, another PhD student of Kirshner's. Phillips had discovered that more luminous supernovae take more time to fade after reaching their peak brightness than do less energetic ones, so there was a relatively simple way to calibrate the stellar explosions. And Riess's "multicolor light-curve shape method" reached yet another level of precision: by carefully looking at the supernova's brightness evolution through different filters, you could even correct for the potential effects of light-absorbing dust.

By January 1998 the two competing teams were ready to present their results at the AAS meeting. The number of remote supernovae was still rather small, and the error bars were still rather large. But the graphs that Perlmutter and High-z member Peter Garnavich showed at the press briefing left no room for doubt. Distant supernovae are not noticeably brighter than you would expect from their redshifts. Ergo, there's not much deceleration going on at all. Certainly not enough to ever stop the universe from expanding.

Interesting enough. But on closer inspection, the graphs told an even more exciting story that Perlmutter and Garnavich didn't pay too much attention to in their presentations. If you knew what to look for, the data seemed to indicate that the most distant supernovae were actually *fainter* than their redshifts would suggest. If true, this implied that the light from a remote supernova took more time to reach us than it would in a coasting universe with a constant expansion rate, not less. In other words, the expansion must have been slower in the past, not faster.

Using different instruments and different algorithms, both the Supernova Cosmology Project and the High-z team came to the same inescapable conclusion: we live in an accelerating universe. Something is speeding up the expansion of space—an outcome so weird that the scientists hardly could believe it themselves, let alone present it as a discovery at the AAS meeting. Could our universe really be that bizarre?

It was only on February 22, at a dark matter conference at the University of California, Los Angeles, that astrophysicist Alex Filippenko was bold enough to claim the discovery of the accelerating universe on behalf of the High-z team, at a time when the Supernova Cosmology Project was still talking about "suggestive hints" and "possible evidence." Three weeks later, the High-z team submitted a thirty-page paper to *The Astronomical Journal,* which ran in the September issue.[5] The final analysis of the Supernova Cosmology Project was completed in the summer of 1998 and published on June 1, 1999, in *The Astrophysical Journal.*[6]

Yes, there have been arguments about priority and credit. Outspoken emails have been exchanged. Responding to the controversy, Kirshner asked a *New York Times* reporter, "Hey, what's the strongest force in the universe?" His own answer: "It's not gravity, it's jealousy."[7] But from a cosmic perspective, the two groups of researchers reached their revolutionary results at exactly the same time. Moreover, the fact that separate teams arrived at the very same conclusion helped to convince doubters and skeptics. It didn't take long before the runaway universe was embraced by astronomers and physicists, taught in university classes, and featured in popular science magazines.

By the time Perlmutter, Schmidt, and Riess shared the 2011 Nobel Prize, most animosities had been forgotten. Cosmologists now have other things to worry about. For dark energy—the ominous name for the mysterious "something" that is accelerating the expansion of the universe—is just as mysterious as dark matter. We know it's there, but we don't know what it is. While it solves a number of problems, it doesn't make our universe any more comprehensible.

Some overconfident astronomers like to boast that we live in an era of precision cosmology. The truth is that we only understand a small sliver of reality. Meanwhile, 95 percent of the universe is one big question mark.

16

Pie in the Sky

Oh, for the good old days.

Some two hundred years ago, the universe was small, simple, and comprehensible. One Sun, seven planets, sixteen moons, a handful of asteroids and comets, maybe a hundred million stars, and a few dozen nebulae. That was it.

Today, just eight generations later, astronomers have catalogued hundreds of thousands of asteroids in our solar system. We know our Sun is just one of a few hundred billion stars in the Milky Way galaxy, which is also home to weird objects like brown dwarfs, pulsars, and X-ray binaries. Most stars have planetary companions; there are more habitable planets in the universe than people on Earth. What's more, our galaxy is just one among hundreds of billions of others, scattered across an expanding cosmos that extends way beyond the reach of our telescopes. The stars in our universe are more numerous than the grains of sand in all of our deserts combined. It's just overwhelming.

And yet, this multitude of galaxies, globular clusters, dust clouds, glowing nebulae, red giants, white dwarfs, planets, supernova remnants, neutron stars, and cosmic debris—this whole material universe of ours—is just the tiny tip of a giant, invisible iceberg. According to current wisdom, we can only see and touch a mere 5 percent of everything there is; most of the universe consists of puzzling dark matter and even more enigmatic dark energy. The solution to this mystery may seem like a cosmological pie in the sky.

Five percent—just one-twentieth of the grand total. If this book represents all of the universe, the familiar baryonic matter would be described on the first fifteen pages, and question marks would fill the rest. Of course, it's impossible to discover that the universe contains less than we already have found. But 95 percent is a lot, especially if no one has a clue as to what we're really talking about.

Still, the concept of dark energy was less surprising to most astronomers and physicists than you might think. To the general public, it came like a bolt from the blue—a seemingly contrived solution to an observational riddle. But scientists like Saul Perlmutter, Brian Schmidt, and Adam Riess knew about the concept all along. Way back in 1917, Albert Einstein toyed with the idea of a "cosmological constant": a mysterious repulsive energy in empty space that would work against gravity. It just took eighty years before astronomical observations suggested that Einstein's hunch may have been spot on after all. For decades, cosmologists had been successful in keeping the cosmological constant out. But when they were confronted with the supernova data, they finally had to give in.

Einstein had a good reason to introduce his "fudge factor," as others have called it. According to the so-called field equations of general relativity, spacetime either had to expand or it had to contract. But at the time Einstein penned his theory, in 1915, he was convinced that the universe at large was static and nonevolving. That's why he inserted in his equations a constant (denoted by the Greek letter lambda, Λ) that enabled this coveted steady state. With the constant in place, the equations described a universe that wouldn't collapse in on itself as a result of its own gravity.

But when Lemaître and Hubble discovered that the universe is expanding, the concept of an unchanging cosmos was off the table. Astronomers realized that the universe is indeed evolving, in perfect agreement with general relativity. Consequently, there was no immediate need for lambda anymore. Einstein once told George Gamow that the introduction of the cosmological constant was the biggest blunder of his career.[1]

However, lambda never really left the stage. For one thing, particle physicists predict the existence of a lambda-like vacuum energy, caused by the constant production and annihilation of virtual pairs of particles and antiparticles in empty space. Although the calculations suggest that this vacuum energy must be incredibly powerful—in stark contradiction with observations—the concept itself doesn't seem to be too far-fetched, at least from a physicist's point of view.

Moreover, a universe with a small but nonzero cosmological constant was welcome news to astrophysicists. In the 1960s and 1970s, they had found stars that appeared to be much older than the universe—which, of course, can never be the case. The cosmological constant might solve this apparent contradiction by beefing up the age of the universe. In a universe with a cosmological constant, the deceleration as a result of gravity is lower, so the universe must have taken more time to slow down to the current expansion rate. Therefore it would be older than a universe in which lambda equals zero, so it could accommodate older stars.

Last but not least, if there is some kind of cosmological constant, this would relieve the pressure associated with the critical density. As we saw in the previous chapter, Alan Guth's 1979 inflationary hypothesis predicts that the universe has a flat, zero-curvature geometry, and a critical density of 10^{-29} grams per cubic centimeter.

If the observed average matter density of the universe turns out to be lower, as it appears to be, maybe vacuum energy can come to the rescue, making up for the shortfall in matter. Remember that according to Einstein's most famous equation, $E = mc^2$, matter and energy are really two sides of the same coin—they both influence the overall properties of spacetime. So a cosmological constant could balance the cosmological ledger even if matter is seemingly missing.

So when astronomers analyzed their observations, and when cosmologists discussed the various theories about the evolution of the universe, they were always careful to state their assumptions about the cosmological constant. "In a matter-only universe," they would write, or "for a model without a cosmological constant," or just "assuming $\Lambda = 0$." Researchers were loath to eject lambda entirely, hence they would acknowledge it even as they discarded it. But most of them disliked the idea—it seemed to be too arbitrary, too complicated.

Others kept a more open mind. For instance, in 1995 Jerry Ostriker and fellow theorist Paul Steinhardt of the University of Pennsylvania published a provocative and visionary paper in *Nature* titled, "The Observational Case for a Low-Density Universe with a Non-Zero Cosmological Constant."[2] Looking at all the available evidence, they concluded that "a Universe having the critical energy density and a large cosmological constant appears to be favored." Mind you, that was more than two years before the supernova data became public. "We would be interested to hear," Ostriker and Steinhardt wrote, "if a serious observational problem can be identified . . . with the flat model which has a substantial cosmological constant. If not, perhaps we have already identified models which, in broad outline, capture the essential properties of the large-scale universe."

So when Perlmutter's Supernova Cosmology Project and Schmidt's High-*z* Supernova Search Team accumulated stronger and stronger evidence for an accelerating universe, the two competing groups did not hit upon something completely unfamiliar. Which is not to say they weren't surprised. Or, as quoted by *Science* magazine, "very excited" (Robert Kirshner), "stunned" (Riess), and "between amazement and horror" (Schmidt).[3] Lambda, that weird and un-explained repulsive property of empty space, turned out to be real after all.

Still, quite a lot of people feel uncomfortable about the idea of an accelerating universe. First there was dark matter—put in by hand, in the words of Jim Peebles, to solve the mystery of galactic rotation curves and other dynamical issues. Now cosmologists are adding yet another ingredient, dark energy, to solve the riddle of supernovae that appear dimmer than expected. To save the appear-ances, some would say. It all smells like epicycles, others objected.

In chapter 12, you read about Mordehai Milgrom, whose theory of modified Newtonian dynamics tries to explain the flat rotation curves of galaxies without invoking dark matter. But in the case of dark energy, the critics do not suggest an alternative explanation for the accelerating universe. Instead they claim that there is no accel-eration at all, and that the supernova hunters are victims of system-atic errors in their observations or analyses.

For instance, Subir Sarkar of the University of Oxford has pointed out that the majority of the supernovae studied by Perlmutter, Schmidt, and their colleagues are in one half of the sky.[4] According to Sarkar, if our Milky Way galaxy happens to be moving in that direction, that would affect the supernova redshifts, leading to er-roneous conclusions. Another objection comes from a Korean-French team led by Yijung Kang and Young-Wook Lee of Yonsei

University in Seoul.[5] They claim to have found evidence that the true luminosity of a type Ia supernova is affected by the age and overall chemical composition of the galaxy in which it occurs. Who knows, type Ia supernova explosions may really have been less luminous in the past than they are now—the past in which the explosions occurred, mind you, since the light we see today was emitted eons ago. "Dark energy might be an artefact of a fragile and false assumption," according to Lee. Still others believe that the light-absorbing and dimming effects of cosmic dust have not been properly taken into account, not even by Riess's elaborate multi-color light-curve shape method.

In a September 2018 interview with *Horizon,* the online research and innovation magazine of the European Union, Sarkar referred to the epicycle pitfall.[6] "The trouble is that people think our standard model of cosmology is simple and it fits the data," he says. "The Ancient Greeks thought the same about Aristotle's model of the universe, in which the sun and the planets revolve around the Earth. But we need to be open to different possibilities. Let's just hope that it doesn't take as long to replace our standard model as it did Aristotle's—2,000 years."

Of course, each and every piece of criticism has been thoroughly analyzed, and in most cases convincingly rebutted, by the dark energy camp. That's how science works. The total number of supernovae on which the conclusions are based has risen to over 700, and the statistical significance of the result has become stronger and stronger. What's more, using the Hubble Space Telescope, Riess and others have identified extremely remote type Ia supernovae that support the dark energy conclusion, given their observed redshifts and luminosities. These explosions happened billions of years ago. Back then, gravitational deceleration was stronger, because the cosmic matter density was higher, while the accelerating effect of dark energy was smaller, simply because there was less space. If you do the math, it turns out that for the first seven or eight billion years after the big bang, dark matter cannot have been the dominant force, and the net acceleration of cosmic expansion had not yet kicked in. The latest graphs of the expansion history of the universe based on supernova observations, which include the Hubble data on the most remote explosions, do indeed bear out this prediction.

Despite all the available evidence, a healthy dose of doubt and skepticism remains. Sarkar, for one, is not convinced. "I believe a lot of cosmological results that support the consensus view came about only because the authors knew beforehand which lamppost to look under," he told *Horizon*. "In other words, they may be suffering from confirmation bias." And in a January 2020 press release, Lee—referring to Carl Sagan's famous mantra that extraordinary claims require extraordinary evidence—is quoted as saying, "I am not sure we have such extraordinary evidence for dark energy."[7]

Anyway, even though the arguments pro and con will probably go on for many years, the vast majority of astrophysicists and cosmologists are convinced by the supernova results. Yes, cosmic expansion started to accelerate a few billion years ago. So yes, there must be some strange kind of dark energy in empty space, which eventually will determine the fate of the universe. But, as Danish physicist Niels Bohr already knew many decades ago, it's difficult to make predictions, especially about the future. As long as no one knows the true nature of dark energy, it's impossible to say anything definitive about how it will behave in the eons to come.

One reason that scientists use the label dark energy—as opposed to the cosmological constant—is that they aren't absolutely sure that the runaway universe is caused by Einstein's fudge factor. ("Dark energy" was proposed by University of Chicago cosmologist Michael Turner.) A cosmological constant would always have the same value, at every point in space and every moment in time. It would be a true, fundamental property of empty space. But dark energy is not necessarily so static; physicists can think (and have thought) of dark energy as an all-pervasive field, somehow comparable to an electrical field or a gravitational field. Provisionally called "quintessence," this field could vary from place to place and evolve over time.

If dark energy is truly a cosmological constant, the universe will keep expanding forever, heading for a cold, dark, empty future. But if instead it's more like quintessence, all predictions are up for grabs. Even a future reversal to an accelerating phase of cosmic contraction cannot be ruled out.

And what if dark energy happens to grow stronger and stronger over time (some physicists like to call this "phantom energy")? In that case, the repulsive effect will eventually be powerful enough to rip everything apart—first galaxies, then stars and planets, then molecules and atoms, and finally elementary particles and spacetime itself. This "big rip" may occur a mere 20 billion years from now, according to an estimate by American astrophysicists Robert Caldwell, Marc Kamionkowski, and Nevin Weinberg in a 2003 paper in *Physical Review Letters*.[8] "It will be necessary," they wrote, "to modify the adopted slogan among cosmic futurologists—'Some say the world will end in fire, Some say in ice'—for a new fate may await our world."

In 1970 Allan Sandage described cosmology as a search for two numbers. Today the field may be better characterized as the science of two mysteries (and many more numbers, by the way). Dark matter and dark energy both play a decisive role in the composition and evolution of the universe. Astronomers measure their effects and include them in their theories. But catchy names and elegant equations do not provide us with deeper understanding. Pretty unsettling, especially if you realize that we're talking about 95 percent of the total mass-energy budget of the universe.

Attempts to strike two birds with one stone have been unsuccessful so far. It would be great if one new, revolutionary insight would solve both riddles at the same time, but as yet, no one has found a way to accomplish that. Nature isn't always kind—Peebles again. Or maybe we just aren't smart enough. Not up to the task. Not yet.

Then again, the discovery of dark energy has a direct effect on our ideas about dark matter. If the universe is flat—a more or less aesthetic assumption at first, but now a consequence of inflation and supported by measurements—it must have the critical density of 10^{-29} grams per cubic centimeter. Since big bang nucleosynthesis tells us that baryons can only account for a small percentage of the critical density, cosmologists had to assume an incredibly large amount of nonbaryonic dark matter—far more than indicated by galaxy and cluster dynamics.

With the advent of dark energy, however, the dark matter hunters could heave a sigh of relief. No longer did dark matter have to account for the flat geometry of the universe all by itself. The new pie chart of the composition of the universe is dominated by dark energy, which carries most of the weight when it comes to flattening spacetime. Whatever its true nature, this puzzling repulsive property of empty space accounts for 68.5 percent of the total mass-energy inventory of the cosmos. Gravitating matter represents only the remaining 31.5 percent.

Which of course doesn't mean we can do without large amounts of dark matter. If we just look at the material content of the universe, it's also dominated by mystery stuff. A whopping 84.4 percent of all the gravitating matter (26.6 percent of total matter) is dark, nonbaryonic, and utterly unfamiliar. The particles that we know of—the building blocks of stars, planets, and people—comprise less than one-sixth (a mere 15.6 percent) of all matter. That's just a 4.9-percent slice of the full cosmic pie.

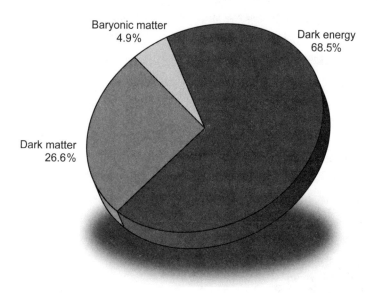

According to the popular ΛCDM model of cosmology, the universe is dominated by mysterious dark energy and dark matter. Just 4.9 percent of the total matter / energy consists of "normal," baryonic matter.

Over the past two decades, the pie diagram of the composition of the universe—68.5 percent dark energy, 26.6 percent dark matter, and 4.9 percent familiar stuff—has become an iconic representation of our cosmic ignorance. The precise percentages have varied a little bit over the years and will probably continue to do so in the near future, but the overall message is loud and clear: our universe is "a riddle, wrapped in a mystery, inside an enigma," in the words of Sir Winston Churchill.

In all its simplicity, the pie diagram also confronts us with our cosmic insignificance. In the sixteenth century, Nicolaus Copernicus told us that Earth is not at the center of the universe. Just over 150 years ago, Charles Darwin made us realize that mankind is not the crown of creation. Now our self-importance receives a third blow. Not only are we nowhere to be found in space and accidental newcomers in time, but the very stuff we're made of is just a minor constituent of the cosmos. It's a humbling message, to say the least.

Homo sapiens is a recent shoot on the evolutionary tree, born a cosmic eyeblink ago on an inconspicuous mote of dust orbiting a run-of-the-mill star in the outskirts of a very average galaxy. So is it too presumptuous to assume that we will one day solve the mysteries of the universe? Maybe. But that shouldn't keep us from trying. We've covered a good deal of distance in the past centuries and decades, and there's every reason to believe that our reach for the skies will yield more answers in the future.

The good old days may not return, but the good new days are likely to be even better.

17

Telltale Patterns

Precision cosmology took off in the jungle of French Guiana, on May 14, 2009, at 10:12 a.m. local time. Watched by dozens of astronomers, technicians, officials, and journalists, a powerful Ariane 5 ECA rocket lifted off from the Guiana Space Center. Rising above the treetops on a pillar of fire and smoke, its thundering roar overwhelmed the songs of birds and the constant buzzing of insects in the humid rain forest. Tucked away in the rocket's nose cone, beneath the Herschel infrared space telescope that took the same ride into space, was the European Space Agency's *Planck* observatory. Its mission: to precisely map the afterglow of the big bang, in search for a better understanding of the universe.[1]

That afterglow, known as the cosmic microwave background (CMB), provides a snapshot of the very early universe, a mere 380,000 years after its explosive origin. Somehow, in the course of 13.8 billion years, minute density fluctuations in this primordial brew evolved into the filamentary distribution of galaxies we see around us today. Studying the "baby photo of the universe," as the CMB has been called, provides information on the cosmic ingredients that shaped this evolution. In fact, *Planck's* measurements of the CMB have convincingly confirmed the dominant roles and the relative contributions of dark energy (Λ) and cold dark matter (CDM). Even without supernova observations and galactic rotation curves, cosmologists can celebrate the ΛCDM model on the basis of *Planck* data alone.

So what exactly is the cosmic microwave background? Let's go back to the beginning to find out. During the first few hundred thousand years of the universe's existence, conditions are too extreme for neutral atoms to form. Instead space is filled with a tremendously hot plasma—a mix of individual protons, neutrons, electrons, neutrinos, and dark matter particles. Like the flame of a candle, this dense plasma is opaque: photons can't freely move throughout space, because they constantly interact with the omnipresent electrons, owing to the latter's electrical charge.

But after some 380,000 years of cosmic expansion, the average temperature of the universe drops below 2,700 degrees Celsius—"cool" enough for nuclei of hydrogen and helium to capture the free electrons. Within just a few tens of thousands of years, all of the plasma is converted into a hot, expanding gas of electrically neutral atoms, and finally the fierce radiation of this blazing inferno can freely stream throughout space, unimpeded by interactions with charged particles.

It's important to realize that every single point in expanding space was once hot enough to produce this glow, almost as bright as the surface of the Sun. The radiation from our own immediate surroundings has long since disappeared into the distance. However, surrounding us on all sides is a shell of space so incredibly remote that its primordial glow is only reaching us today. During this 13.8-billion-year trip, the energetic radiation got redshifted by the expansion of the universe, and by the time it reaches our detectors, little is left beyond a cold, faint, almost imperceptible hiss at radio wavelengths. As described in chapter 1, it was this faint cosmic microwave background that Bell Labs radio engineers Arno Penzias and Robert Wilson accidentally discovered in 1964.

The big horn antenna used by Penzias and Wilson detected the same amount of radiation no matter which way it was pointing. But right from the start, it was clear that the cosmic microwave background could not—and should not—be perfectly smooth across the sky. The very existence of galaxies and clusters in the present universe tells you that there must have been tiny density fluctuations in the primordial soup. These small variations in density should be discernible as equally tiny patches of slightly higher and slightly lower temperature in the background radiation.

Building on pioneering 1946 work by Soviet physicist Evgeny Lifshitz, University of Texas researchers Rainer Sachs and Arthur Wolfe were among the first to make quantitative predictions about the expected size of these variations. In their 1967 paper in *The Astrophysical Journal,* Sachs and Wolfe concluded, "We have estimated that anisotropies of order 1 percent should occur in the microwave radiation if this radiation is cosmological"—something that was not generally accepted at the time.[2] But despite ever more precise observations in the 1970s and 1980s, the predicted temperature variations—the anisotropies in question—were not found. The CMB turned out to be improbably smooth, which is what led Jim Peebles to propose the existence of nonbaryonic cold dark matter in 1982.

Improbably smooth, but not perfectly smooth. The big discovery came in January 1990, at the American Astronomical Society meeting in Washington, DC, where scientists presented the first results of NASA's Cosmic Background Explorer (COBE) satellite. COBE had been mapping the cosmic microwave background at an unprecedented sensitivity since its launch in November 1989. In the first few weeks of its operational lifetime, it precisely measured the

CMB's spectrum. It also detected the long-awaited temperature fluctuations, albeit at a much smaller level than Sachs and Wolfe had predicted twenty-four years earlier.

While the average temperature of the microwave background is 2.725 kelvin (just a few degrees above absolute zero), the "hot" and "cold" patches were off by no more than 30 millionths of a degree— an anisotropy not of 1 percent but 1 / 1000 of 1 percent. Finally, cosmologists had numbers to put their hands on. COBE principal investigators John Mather and George Smoot received the 2006 Nobel Prize for their breakthrough achievements.

However, COBE's angular resolution wasn't terribly good, meaning its view of the CMB was still a bit blurry. The satellite's instruments did register small temperature variations all over the sky, but they couldn't discern the smallest hot and cold spots—akin to you or me looking at a pointillistic Seurat painting from afar, unable to make out the individual colored dots. But by the late 1990s, high-altitude balloon experiments succeeded in resolving the small-scale structure of the CMB, albeit in relatively small areas of sky. And in June 2001, NASA launched the successor to COBE: the Microwave Anisotropy Probe. After David Wilkinson, a member of the mission's science team, died in 2002, the satellite was rechristened the Wilkinson Microwave Anisotropy Probe.

Between 2009 and 2013, ESA's *Planck* spacecraft (named after the famous German physicist Max Planck) took the CMB measurements to yet a new level of sensitivity and precision. *Planck's* map of the microwave background was presented to the world on March 21, 2013, at a press conference at the ESA Headquarters in Paris, but it took scientists another five years to complete the data analysis. The results were published on July 17, 2018, in a series of twelve papers in *Astronomy & Astrophysics*.[3] According to project scientist Jan Tauber, "This is the most important legacy of *Planck*. So

Minute temperature variations in the cosmic microwave background radiation reveal tiny density fluctuations in the very early universe, just 380,000 years after the big bang.

far, the standard model of cosmology has survived all the tests, and *Planck* has made the measurements that show it."

That is a big claim. What Tauber was saying was that the increasingly precise CMB findings revealed that we live in a flat universe, just as predicted by Alan Guth's 1979 inflationary hypothesis described in chapter 15. The measurements also confirmed the existence of dark energy.

How does the baby photo of the universe tell us about its composition and evolution? It all has to do with the intimate coupling of baryonic matter and radiation during the first 380,000 years of cosmic history. Shortly after the big bang, the hot primordial plasma wasn't perfectly smooth, probably because of initial quantum fluctuations that literally got blown up to macroscopic dimensions by the exponential expansion of inflation. As a result, the newborn universe was littered with small regions that had a slightly higher-than-average density: more protons, neutrons, and electrons ("normal" matter) but also more dark matter particles. And while the dark matter particles responded only to gravity, the baryonic plasma also strongly interacted with the omnipresent high-energy photons.

Because of gravity, an overdense region wants to contract even further, becoming denser still. But as the density grows, radiation pressure also increases, causing the clump to expand instead—at least as far as the baryons are considered. The result, in the early universe, was a propagating longitudinal wave of higher and lower pressure in the baryon-photon fluid, very similar to a sound wave in air, albeit with a much larger wavelength. Meanwhile, the central dark matter overdensity remained where it had been all the time.

If the early universe only had one density anomaly, it would be easy to recognize the resulting sound wave pattern. In reality, however, there was a cacophony of sound waves of varying wavelengths and amplitudes, racing through the expanding cosmic plasma in each and every direction at a velocity of almost 60 percent of the speed of light. If you like poetic metaphors, you could call it the birth cry of the cosmos.

This primordial soundscape lasted as long as baryons and photons were strongly coupled. But as the powerful interaction between matter and radiation came to a halt some 380,000 years after the birth of the universe, the baryonic sound waves were quite suddenly silenced. At this point the photons started to freeride across the universe in the form of the cosmic microwave background, while the baryons found themselves in a three-dimensional distribution of higher and lower density—a freeze-frame imprint of the hodge-podge of acoustic oscillations.

The speckled temperature pattern of the cosmic microwave background is directly related to this primordial density distribution. Thus cosmologists can use the *Planck* map of the CMB to study these baryon acoustic oscillations, as they're called. It's a bit like reconstructing the various sounds in a noisy environment from studying a very detailed, short-exposure photograph of a human eardrum. The eardrum is vibrating at many different frequencies and amplitudes at once, but by analyzing the complicated freeze-frame pattern of oscillations, it's possible to tease out the individual sound waves. Likewise, a detailed analysis of the CMB pattern tells you the amplitude (power) of the radiation's numerous constituent waves as a function of wavelength. The resulting graph is known as the CMB power spectrum.

The precise sound wave pattern in the 380,000-year-old universe—at the time the cosmic background radiation was released—is determined by a surprisingly small number of variables. Particularly important are the density of baryons, the density of nonbaryonic particles, and the so-called sound horizon: the distance a sound wave in the expanding plasma could travel in the time preceding the decoupling of matter and radiation. It turns out that changing one of these variables by even a relatively small amount will affect the precise shape of the CMB power spectrum. By turning the equations around, so to say, you can start out with the observed power spectrum and work your way back to derive the baryon and dark matter density, as well as the sound horizon, at the time of decoupling.

It shouldn't be surprising that there's a close relationship between wave power and sound horizon. Quite the same is true for organ pipes: the length of the pipe (the distance a sound wave can travel) determines which wavelengths have the largest amplitudes— remember how Peebles produced two very different sounds by blowing air over two empty plastic bottles of different sizes, as described at the very beginning of this book. Of course organ pipes are not expanding, and gravity doesn't play a role during a Bach concert, but the principle is the same: every size has its own favored frequencies—its own set of ground tone and corresponding overtones.

Now it gets interesting. The sound horizon at the time of decoupling turns out to be on the order of 450,000 light-years—that's the distance covered by the baryon acoustic oscillations after 380,000 years. (No, there's no contradiction here: the oscillations travel at almost 60 percent of the speed of light, but thanks to cosmic expansion, they end up almost twice as far from their starting point as they would in a static universe.) So at the time the oscillations come to a halt, every original overdensity in the newborn universe was surrounded by a spherical shell of higher-than-average density with a radius of 450,000 light-years.

This radius shows up as a preferred distance in the pattern of hot and cold spots in the cosmic microwave background, intimately related to the location of the first "peak" in the CMB power spectrum. Because of the many overlapping density waves in the primordial plasma, temperature anisotropies looked random at first, but if you measure the distance between every possible pair of spots all over the sky, a pattern emerges: the number of pairs that are 450,000 light-years away from each other is significantly higher than you would expect from a random distribution.

This brings us to the proof of a flat universe. When studying the statistical distribution of the CMB's hot and cold spots, astronomers don't measure the distances separating them in light-years but in terms of the angle they subtend in the sky. What they find is a preferred angular distance of about one degree. This preferred angular distance should correspond to the preferred physical distance of 450,000 light-years (remember that the CMB photons came from very far away—they took some 13.8 billion years to arrive on Earth—so it's not that surprising that a separation of 450,000 light-years only subtends one degree in the sky).

But if the background radiation traveled through a "closed" universe, with an overall positive curvature, two points separated by 450,000 light-years would subtend an angle of *more* than one degree in our sky. In an "open" universe, with a negative curvature, the angle would be *smaller* than one degree. Only in a flat, Euclidean universe with no overall curvature does a linear distance of 450,000 light-years at 13.8 billion years ago correspond to an angular distance of one degree in the sky today.

Thus the telltale patterns in the cosmic microwave background have revealed fundamental properties of the cosmos. In agreement with inflationary theory, we live in a flat universe, which means that the total mass-energy density must equal the critical density. From the *Planck* data, it also follows that the baryon density is just 4.9 percent of the critical value, while the density of nonbaryonic cold dark matter makes up another 26.6 percent. Consequently, the remaining 68.5 percent of the mass-energy budget of the universe must be in the form of dark energy.

The remarkable thing here is that these values are completely independent of earlier astrophysical estimates. No rotation curves of galaxies, no cluster dynamics, no type Ia supernova measurements—a detailed map of the cosmic microwave background is all you need to conclude that our universe is dominated by dark energy and dark matter. As the *Planck* team modestly noted in the abstract of their 2018 paper on cosmological parameters, "We find good consistency with the standard . . . ΛCDM cosmology."[4]

So astronomers have a pretty good picture of what the universe looked like just 380,000 years after the big bang, and a detailed look at the CMB baby photo reveals a lot about the most fundamental properties of the cosmos. Then again, it's only a baby photo—a glimpse of the newborn universe. If the present-day universe is a

fifty-year-old woman, the map of the cosmic microwave background shows her features when she was just half a day old. How
did the baby grow and develop into the adult?

Well, for starters, the universe has expanded by a factor of almost
1,100 since the CMB was released (if this were true for the human
baby, our fifty-year-old grown-up would be 550 meters tall). Given
this impressive growth spurt, you would expect the highs and lows
in the density distribution of the early universe to have smoothed
out over time, but in fact the gap between them became ever more
pronounced as a result of gravity. At every point in space, the matter
density dropped as a result of cosmic expansion, but in overdense
regions the density decreased much more slowly than in underdense
regions, so that the density contrast increased. This process has been
studied in detail with the help of massive computer simulations, as
we saw in chapter 11.

As the universe aged, overdensities in the distribution of dark
matter—which had been able to grow during the first 380,000 years
of cosmic history, since dark matter wasn't interacting with
radiation—continued to attract even more matter, both nonbaryonic and baryonic. But the same was true for the slightly overdense
shells surrounding these dark matter concentrations at 450,000 light-
years distance—the crests of the baryon acoustic oscillations that
"froze" at the time of decoupling. They, too, began to gravitationally attract more and more dark and "normal" matter.

Eventually, this intricate pattern of density variations evolved into
the filamentary large-scale structure of the present universe, com-
monly known as the cosmic web. So by carefully looking at the
spatial distribution of galaxies, astronomers should still be able to
recognize the preferred distance of 450,000 light-years that was
found in the map of the cosmic microwave background, although

by now, it has expanded to some 500 million light-years. Even after 13.8 billion years of cosmic evolution, the frozen-in oscillations must still be visible out there.

Of course, photos of the night sky do not show conspicuous circular arrangements of galaxies, as some misleading illustrations in popular stories about baryon acoustic oscillations seem to suggest. Remember that we're talking about a pretty subtle effect, imprinted on top of a much smoother galaxy distribution. But if you made a three-dimensional map of tens of thousands of galaxies and measured the physical distance between every possible pair, you would expect that this so-called two-point correlation function shows a bump at a separation of 500 million light-years—at least in the present local universe. At much larger distances, where we look further back in time, the oscillations should be correspondingly smaller, since the universe hadn't yet expanded to its present dimensions.

It wasn't until 2005 that galaxy surveys were large and deep enough to convincingly reveal this telltale pattern. By that time, both the Two Degree Field Galaxy Redshift Survey and the Sloan Digital Sky Survey—two programs that you read about in chapter 6—had finally succeeded in detecting baryon acoustic oscillations in the 3D distribution of remote galaxies.[5] Here was a clear link between the characteristics of the pointillistic CMB map and the statistical properties of the distribution of galaxies: an unmistakable connection between the baby photo and the grown-up lady. The pieces of the cosmological puzzle started to fit together, and it all made sense.

The central message is that the past is visible in the present. Powerful sound waves that propagated through the blazingly hot plasma in the first few hundred thousand years of cosmic history left their mark on the large-scale structure of the universe—a characteristic birth mark that is also recognizable on the speckled baby photo provided by studies of the cosmic microwave background.

By the time I saw the *Planck* observatory lift off into space—the mission that would yield the most precise cosmological parameters to date—there was hardly any cosmologist left who doubted the ΛCDM model. Evidence for the existence of dark matter had piled up slowly but surely over the decades. Dark energy, despite being a relative newcomer on the theoretical stage, was also impossible to ignore by 2009. ΛCDM provided a consistent description of the composition and evolution of the cosmos that perfectly fit all the observational data, as well as the latest supercomputer simulations. No other model of the universe could claim the same incredible level of success.

Still, I couldn't suppress those nagging "what if" thoughts. What if cosmologists were chasing a mirage? After all, every single piece of evidence for dark matter and dark energy is circumstantial. No one has ever detected a dark matter particle. No one has ever directly measured the accelerating expansion of space. All we have is indirect proof. Galaxy dynamics. Gravitational lensing measurements. Supernova observations. Baryon acoustic oscillations. What if they are all leading us astray? Could we be busy painting ourselves into a corner, like the nineteenth-century ether believers we met in chapter 1? What if dark matter and dark energy are nothing more than smart mathematical tricks to explain away our fundamental ignorance—the epicycles of modern physics?

Seeing the powerful rocket with its precious payload veer off into space, I wondered where cosmology as a science was going. Larger numbers of more precise measurements—great. But if ΛCDM is true, if the familiar world around us makes up a mere 4.9 percent of a largely mysterious universe, isn't it about time to change at least one of the big remaining question marks into an exclamation mark? To really lay our experimental hands on dark matter, at the very least?

On the other side of the Atlantic, not high up in space but deep underground in the Italian Apennines, ambitious physicists had exactly the same thought. It was about time.

Their trump card?

Xenon.

PART III

Trunk

.

18

The Xenon Wars

Like an army of pitchforks, Manhattan's skyscrapers seem to attack the dark clouds descending on New York City. Snow is expected later in the day, but right now the Empire State Building, the Chrysler Building, 432 Park Avenue, and the Freedom Tower occasionally catch a brief shaft of low January sunlight.

"It's a marvelous view," says Elena Aprile, with an amiable Italian accent. "I never get tired of it." We are meeting in the sky lounge on the forty-sixth floor of her Brooklyn apartment building.[1] After the interview, she offers me a really good espresso in her stylish kitchen, showing me a photo of her first grandchild—"It's so great to see your own daughter in the role of a mum." And she shows me another photo, of herself in the late 1970s, at age twenty-three. There she is at CERN, young and ambitious—her own words.

As the founder and long-time spokesperson of the XENON dark matter experiment, her ambitions are still sky-high.[2] Aprile has always wanted to do better than others. No, she hasn't found dark matter. Not yet. But it could happen any time, and it had better be her new XENONnT detector that makes the Nobel Prize–worthy discovery.

For Aprile is not the only one dreaming of fortune, fame, and Stockholm. In the United States and in China, other groups are battling for the same breakthrough, using the same liquid-xenon technology. Former collaborators. People trained by her. Even her own ex-husband. If this is a war, she is determined to win. To be the best, as always.

Quite a character, as XENONnT's technical coordinator Auke Pieter Colijn told me when I visited the Laboratori Nazionali del Gran Sasso.

Born in Milan, young and curious Elena Aprile studied physics at the University of Naples. During her third year, she applied for a summer opportunity at CERN. Getting selected and assigned to the research group of Carlo Rubbia was the best thing that had happened in her life to that point, even though Rubbia was a pretty intimidating person, especially toward women. And that was in May 1977, years before he would become a Nobel laureate and CERN's director-general.

Apart from these power and gender issues, CERN was a scientific paradise, and a gateway to international physics. Aprile never really went back to Italy—or to her boyfriend, for that matter. She stayed with Rubbia for more than half a year and met German physicist Karl Giboni, whom she married in 1981 during her PhD study at the University of Geneva.

In 1983 Rubbia offered both Aprile and Giboni postdoc positions in his Harvard University research group, working on an underground experiment to study the possible decay of protons. Slowly, she worked her way into the male-dominated community. But working for Rubbia was exhausting. After he shared the 1984 Nobel Prize for the discovery of W and Z bosons, he became even harder to deal with. He would fly in from Europe—people sometimes joked that Rubbia spent half of his life in business class—have dinner with Aprile and Giboni, tell them what they did wrong and where they failed, make them feel bad, and fly back to Geneva. By late 1985, Giboni was fed up. "I quit," he told Rubbia, before accepting a research position at a company in New York. Aprile left Harvard in January 1986 to join the physics department of Columbia University.

Columbia was where her love for liquid noble gases developed. First she built argon-based neutrino detectors. Next came a balloon-borne gamma-ray telescope that used xenon as the detector liquid. And in 2001, when she started to look for projects with better funding prospects, she became intrigued by a British experiment that used liquid xenon to hunt for dark matter.

The British ZEPLIN experiment (Zoned Proportional scintillation in Liquid Noble gases—yet another unwieldy acronym) had been proposed by the UK Dark Matter Collaboration. At 1,100 meters underground in the Boulby potash mine in Northeast England, shielded from the unabated bombardment of cosmic rays from above, scientists meticulously monitored a small container filled with one liter (about three kilograms) of ultra-cold liquid xenon. The goal was to detect extremely rare interactions between atomic nuclei and WIMPs—the weakly interacting massive particles (see chapter 10) that were thought to constitute the mysterious, invisible stuff accounting for most of the gravitating mass in the universe. A newer, larger and more sensitive version of the detector was under construction, and Imperial College physicists even had ZEPLIN-III in the pipeline. Clearly here was something to compete with, something to defeat.

At a big physics conference in Aspen, Colorado, in the summer of 2001, Aprile acquainted herself with the topic of dark matter, which she didn't know too much about at the time. That same year, she wrote a proposal for funding from the National Science Foundation to develop her own detector.

The money came through; Aprile was now in charge of her own dark matter–detection experiment, the XENON program. Which, incidentally, wasn't healthy for her personal life. In 1996 Giboni had

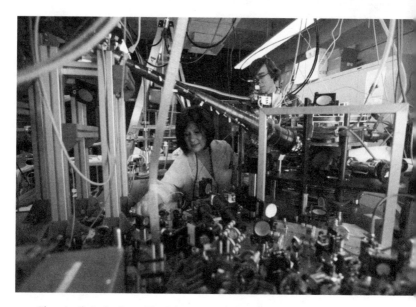

Elena Aprile in the Atom Trap Laboratory at Columbia University, testing technology to trap krypton atoms that contaminate liquid xenon used in dark matter detection experiments.

ended his ten-year flirtation with industry, joining his wife at Columbia as a senior research scientist. By 2001 Aprile was a full professor at Columbia, and Giboni was still working for her. For the XENON project, their collaboration was most fruitful, but for their marriage, it wasn't. It was gender and power all over again.

"I've learned you can't win without losing something else," Aprile says while staring out of the sky lounge's large windows. Across the East River, the tops of the tallest skyscrapers begin to disappear in the gray clouds. "I've been successful in my work, but not in my personal life. I lost my husband in this process."

Two other XENON team members were Brown University physicist Richard Gaitskell and his Princeton colleague Thomas Shutt. Gaitskell and Shutt had worked together on the Cryogenic

Dark Matter Search, an experiment that uses semiconductor detectors (more on those later). But Gaitskell and Shutt were excited by the prospects of liquid xenon.

After building a three-kilogram proof-of-concept detector called XENON3, the plan was to develop the most sensitive liquid-xenon dark matter experiment in the world, leapfrogging the British competition. Of course, just like ZEPLIN, the new detector had to be shielded from cosmic rays, so Aprile, Giboni, Gaitskell, and Shutt set out on a scouting trip for a suitable underground physics laboratory—in other words, for a deep enough mine.

The old Creighton nickel mine near Sudbury, Ontario, was one possibility. Already home to a neutrino physics laboratory called SNOLAB, Creighton was over two kilometers deep and relatively close to the US East Coast, where the assorted team members were based.[3] Meanwhile, the cryogenic experiment that Gaitskell and Shutt had been working on was based at the Soudan iron mine in Minnesota—farther away, but familiar ground. The Homestake gold mine in Lead, South Dakota, was another candidate; a neutrino experiment had been running there since the late 1960s. Then there was the Waste Isolation Pilot Plant in New Mexico, a deep geological repository for radioactive waste. And why not go to Europe? Even if you were competing with the ZEPLIN team, you could base your experiment in the same Boulby mine at the edge of the North York Moors. Finally, in Aprile's home country of Italy, Shutt was working on the Borexino neutrino experiment in the Gran Sasso tunnel—a site well-known to Aprile and Giboni because of their involvement with ICARUS, another neutrino experiment, initiated by Carlo Rubbia.[4]

In the end, they chose the Gran Sasso lab for a number of practical reasons—and yes, they also joked about culture, food, and climate. Aprile's National Science Foundation money and Gaitskell's

Department of Energy funding enabled the rapid construction of XENON10—a name that stuck, although the experiment eventually contained fifteen kilograms of liquid xenon in a cylindrical vessel slightly larger than a one-gallon paint can. The detector was built at Columbia's Nevis Laboratory, shipped to Italy, and installed in March 2006. By that time, the collaboration had grown to over thirty people. Data collection began that same year.

The development and deployment of XENON10 had been a frantic, rollercoaster experience. The team worked eighteen-hour days while keeping a low profile—below the radar of the general physics community. So the first results, announced in 2007 and published in *Physical Review Letters* in January 2008, took the world by surprise.[5] No, XENON10 did not detect WIMPs—it didn't detect anything unexpected. But, almost overnight, it was by far the most sensitive dark matter experiment ever constructed. As such, it provided important new upper limits for the WIMP interaction rate, constraining theoretical models that had never been experimentally tested before.

To understand the expected WIMP interaction rate, you have to realize that the Sun and the Earth, in their 250-million-year orbit around the center of the Milky Way galaxy, plow through a more-or-less static halo of dark matter particles at a velocity of some 220 kilometers per second, or almost 800,000 kilometers per hour. If dark matter consists of WIMPs, and if each WIMP is about a hundred times more massive than a proton, you'd expect about one dark matter particle in a volume as large as a Rubik's Cube. But thanks to their relative velocity, almost a billion WIMPs pass through your body each and every second.

WIMPs don't feel the electromagnetic force, so they don't interact with electrons. They do feel the weak nuclear force, however, and very sporadically, they are expected to collide with atomic nuclei and interact with the constituent quarks. To detect this interaction, you need to closely watch a large number of nuclei, remove—or at least recognize—every possible disturbing background signal, and patiently wait. Liquid xenon (at a temperature of −95°C) turns out to be a perfect target material for detecting such collisions, as it exhibits almost no natural radioactivity, which would ruin the observations.

Here's how the actual detection works. When a xenon nucleus gets hit by a WIMP, it receives a jolt. In a tiny area, xenon atoms lose some of their electrons as a result (a process known as ionization) and briefly enter an excited, molecule-like state. When everything returns to normal again, a faint flash of ultraviolet light, known as a scintillation signal and lasting no more than 20 nanoseconds or so, is emitted at a wavelength of 178 nanometers. This diminutive signal can be registered by photomultiplier tubes mounted in the top and bottom of the cylindrical xenon vessel, each tube sensitive enough to detect a single photon.

The biggest problem is that more mundane (and more frequent) interactions produce a similar excitation and a similar scintillation signal, at the same wavelength. Yes, experimentalists do what they can to shield the detector from cosmic rays and to purify the xenon, as we already saw in chapter 2. But nothing is ever perfect, and you never get rid of all the unwanted background signals.

Because of the self-shielding capability of liquid xenon against these background signals, WIMP hunters are especially interested in scintillation from the core of their target: a signal from the central region of the xenon container is less likely to be a background event than is a signal from the edge. But just registering a brief flash of UV light doesn't give you this valuable information. Therefore Aprile and her team built a dual-phase detector, a design pioneered in ZEPLIN-II.

"Dual phase" refers to the fact that xenon exists both in the liquid phase and in the more common gas phase. In a dual-phase detector, a second signal is produced in the thin layer of gaseous xenon on top of the liquid. As we've seen, xenon nuclei lose some of their electrons as a result of a WIMP interaction. Impelled by a strong electric field erected around the detector, these negatively charged electrons drift vertically upward from the interaction point at a speed of about two kilometers per second. When the electrons reach the interface between the liquid and the gaseous xenon, they are extracted and accelerated by an even stronger electric field, resulting in a one-microsecond burst of electroluminescence in the gas—the same process that makes a neon sign glow.

So every interaction in the xenon container produces two distinct signals: a very brief flash of scintillation light at the instant the interacting particle strikes the xenon, followed by a more prolonged electroluminescence. The time delay between the two signals tells you the depth in the detector at which the interaction took place. Combining that information with the number of photons observed by each of the photomultiplier tubes in the top and bottom of the detector gives you the coveted three-dimensional position of the interaction. Moreover, the relative strengths of the two signals further helps you to discriminate between WIMP interactions—if they occur at all—and events caused by background beta particles or gamma rays.

If it sounds complicated, that's because it is. But experimental physicists like challenges, and what could be more exciting and rewarding than designing and building the most sensitive detectors, with the aim of solving nature's best-kept secrets? At least, that urge to understand the physical world is what has motivated Gaitskell ever since he was a child. When little Richard was eight years old, his mother found him sitting naked in the tub, drawing lines on the bathroom tiles with a permanent marker and trying to calculate the trajectory of a jet of water.

Just weeks before I visited Gaitskell in his office in Providence, Rhode Island, he had confronted another mystery of physics while skiing in Utah: gravity.[6] With his broken leg resting on a small swivel stool, he fetches me coffee and reaches for a tiny box, almost hidden beneath stacks of papers and magazines. Out comes a twelve-gram rectangular piece of ultra-high-purity niobium crystal the size of a playing card. "That's what started it all," he says.

Gaitskell's first job had nothing to do with particle hunting: after obtaining a master's degree in physics from Oxford, the Englishman worked as an investment banker at Morgan Grenfell in London for four years. In 1989, when he concluded that economics was not intellectually challenging enough for him, he returned to Oxford and worked on niobium crystals with his thesis advisor, Norman Booth. In 1995, two years after receiving his PhD, he moved to the Center for Particle Astrophysics at the University of California, Berkeley, the epicenter of dark matter detection at the time.

The connection here is semiconductors, of which niobium is one example. Like liquid xenon, semiconductor crystals cooled down to just a few thousandths of a degree above absolute zero can be used to detect dark matter. In theory, a WIMP passing through the crystal might hit a nucleus along the way. Extremely sensitive superconducting detectors attached to the crystal could register the resulting vibration and charge displacement. Envelop the cryogenic device cooling the crystal in thick layers of lead to keep out as much natural radioactivity out as possible, take the whole thing into a deep mine to shield it from cosmic rays, and you're in business.

As an expert in semiconducting crystals, Gaitskell went to Berkeley to work with Bernard Sadoulet's Cryogenic Dark Matter Search group, which was using stacks of hockey puck–sized germanium and silicon crystals—the same semiconductor materials that are used in computer chip technology and solar cells.[7] Gaitskell spent a couple of years in clean rooms working with these crystals, testing them in the shallow Stanford Underground Facility on the other side of the San Francisco Bay, before deciding to change course. It just took too much effort, too much manual labor, to make these detectors significantly larger, which was the only way to make them more sensitive. So when he moved to Brown University in 2001, he got in touch with Aprile at Columbia.

"I realized this wasn't going to be a sprint," he says, "but more like a marathon, where every new mile is harder than the previous one. For each stage, you need better legs, meaning bigger detectors." Working with xenon offered that possibility. Gaitskell knew all about dark matter; Aprile knew all about liquid noble gases—it seemed like a perfect match.

Except it wasn't. In 2007, after completing XENON10 and during the design phase of its successor XENON100, the collaboration fell apart. Or, rather, it exploded. Gaitskell and Aprile were two equally strong and ambitious personalities. They disagreed on nearly everything, including the best way to run a large international project—in particular, the desirability of moving the experiment back to the United States.

Thanks to a $70 million grant from South Dakota businessman and philanthropist Denny Sanford, an old plan to turn the Homestake gold mine in South Dakota's Black Hills into a physics laboratory could finally be realized. By 2007 Gaitskell and Shutt wanted to build the next liquid-xenon detector in the mine. Four of the seven US teams within the XENON collaboration agreed—working "nearby" in your home country would be much more efficient than commuting to and from Europe. Aprile, however, insisted on staying in Italy. XENON100, eventually containing 165 kilograms of liquid xenon, was under construction. The Gran Sasso lab was readily available, and she didn't want to lose momentum. She wanted to be the first—and the best. The direct detection of dark matter could be right around the corner; it would be madness to lose any time.

So Gaitskell and Shutt moved on and began developing their own Large Underground Xenon dark matter experiment (LUX) at the brand new Sanford Underground Research Facility.[8] Eventually they assembled a group of over a hundred physicists from twenty-seven institutions. Yes, they lost two or three years in the process, but LUX was more sensitive than XENON100, sporting 370 kilograms (more than 100 liters) of target material, shielded by 260,000 liters of water to keep unwanted neutrons out. Construction began in 2009, deployment at the 1,480 meter deep Sanford lab occurred in 2012, and the first data were taken in 2013, just one year after Aprile's team published new limits on the WIMP interaction rate derived from XENON100 findings.[9] "It all worked very well," says Gaitskell. "We blew everyone out of the water."

Now the race was really on. In 2014 the XENON collaboration, which had attracted new research groups from various European countries, started building a yet a larger detector. A xenon vessel the size of a washing machine, containing a whopping 3.2 tons of the liquid stuff. No fewer than 248 photomultiplier tubes. A water tank almost as tall as a three-story building, with a volume of 700 cubic meters. And an aggressive schedule, with the first data collected in 2016, the year in which LUX was decommissioned. XENON1T, as the new detector was called, had its first results published in May 2017.[10]

And it wasn't a war between two armies anymore. In 2009 Xiangdong Ji of the University of Maryland, a former collaborator on XENON100, started working on a competing program in the Far East, where China wanted to run its own flagship experiment.

Aprile's ex-husband Giboni accepted a professorship at Shanghai Jiao Tong University to join the project, which only strengthened her determination.

Known as PandaX, for Particle and Astrophysical Xenon detector, the Chinese instrument is located beneath 2,400 meters of rock, mostly marble, at the Jinping Laboratory in Sichuan Province.[11] It's not only the deepest but also the "quietest" underground physics lab in the world. The first PandaX detector contained 120 kilograms of xenon; PandaX-II, which started operations in March 2015, increased this volume by a factor of four, beating the sensitivity of LUX. Eventually, the Chinese team hopes to build a 30-ton detector.

"Competition is a good thing," Gaitskell says. "People work harder when they know there are others." Unwilling to give up, working harder was exactly what he did. With LUX surpassed by both XENON1T and PandaX-II, he joined forces with the British ZEPLIN group. Just before Christmas 2019, the various parts of the 10-ton LUX-ZEPLIN detector were moved underground in South Dakota. As of this writing, operations are about to commence.[12] Meanwhile, as we saw in chapter 2, Aprile's Gran Sasso team has decommissioned XENON1T and completed construction of XENONnT, which is very much comparable to LUX-ZEPLIN in terms of sensitivity, even though the target mass is a bit lower at 8.6 tons.

The battle isn't over yet, and it probably won't be for many years to come. Nevertheless, Gaitskell says, "I'm confident that dark matter will be found, somehow, someday. Otherwise, I wouldn't do this. I want to answer that question." Not an easy task, he adds, as he shifts his broken leg to a somewhat more comfortable position on the swivel stool. "It may take much longer than I'm hoping for. For who's to say that this problem is to be solved by a single generation?"

In New York City, Elena Aprile is dreaming about dark matter and about a Nobel Prize–winning discovery all the time—at least, that's what she tells me, after showing that black-and-white photo of her ambitious twenty-three-year-old self. Finding a signal—how amazing and awesome would that be. Listening to her stories and looking at her strong-willed face, I can't help believing her. This woman's whole life revolves around the dream of detecting dark matter and the desire to make it come true. It could happen any time now.

As I cross the East River back into Manhattan, it has finally started to snow. The sidewalks are slippery; most of the high-rise buildings are lost from view. Walking through the snowfall and leaving footprints in the fresh white carpet, I try to imagine—without much success—that a billion invisible WIMPs pass through my body every single second, day after day, year after year. Through the Brooklyn Bridge. Through the Freedom Tower. Through our planet. Through the most sensitive detectors in the world, in deep underground laboratories.

The mysterious stuff that governs the large-scale behavior of our universe is all around us, but so far we haven't been able to detect it. And it's not for lack of determination, effort, and perseverance. So are physicists chasing phantoms after all? Or maybe the beast really is out there, but scientists are looking for the wrong one, using the wrong equipment?

It's time to take a closer look at some of the other instruments in the experimentalists' toolbox. And what about that other Gran Sasso experiment—the one that does claim to see evidence for dark matter?

19

Catching the Wind

Rita Bernabei doesn't want to talk to me on the phone.

Which is remarkable for someone who claims to have detected dark matter.

It's nothing personal—at least not that I'm aware of. It's just that the Italian physicist doesn't do oral interviews with journalists. "It is our general policy to answer to written questions in written form," she informs me in an email. "We think that this behavior assures better transparency for both the journalists and the collaboration."[1]

I'm not so sure.

When Auke Pieter Colijn walked me through the Laboratori Nazionali del Gran Sasso, we passed the lab space of Bernabei's dark matter experiment, which is simply called DAMA. The door was locked, and there was no one around. "It's a very closed community," Colijn told me. "I don't know the DAMA physicists personally."

For more than twenty years, Bernabei's team has been studying particle interactions in high-purity crystals of sodium iodide. And for more than twenty years, the researchers have claimed to see an annual variation in the event rate, with a few percent more interactions occurring in early June and a few percent less in early December. Time after time, year in, year out.

Such an annual modulation is not something you would intuitively expect from any known background source, be it cosmic rays, muon-induced neutrons, or beta particles and gamma rays from natural radioactivity. These particles leave their mark at an unchanging rate. However, annual variation is to be expected from WIMP-like dark matter particles. As we saw in the previous chapter, our solar system moves through the Milky Way's dark matter halo at a speed of some 220 kilometers per second. Our home planet of course orbits the Sun once per year, at a much lower velocity of just under 30 kilometers per second, and it just so happens that in early June, the two velocities more or less add up, increasing the dark matter "wind speed" that we experience here on Earth. Consequently, the number of dark matter interactions in a terrestrial detector should be a bit higher than average during that time. In early December, Earth moves in the opposite direction, and we should measure a slightly lower interaction rate.

So here's the bottom line: DAMA claims to see this annual variation, but mainly because they're the only one, no one else believes that their results are convincing proof of the detection of dark matter. Oh, and they shouldn't be that secretive—that certainly hasn't helped either.

Efforts to directly detect dark matter particles only started in the mid-1980s, after Harvard postdoc Katherine Freese ran into Polish physicist Andrzej Drukier at a conference in Jerusalem. Drukier had been working on technologies for detecting neutrinos from the Sun, and physicists had realized that a similar approach might offer a chance of looking for WIMPs—the most popular dark matter candidate.[2]

Through the weak force, WIMPs in the galactic halo would interact with atomic nuclei every now and then, and working from a few assumptions about WIMP properties, you could easily calculate the expected event rate for a variety of cold dark matter candidate particles. Together with David Spergel, who was a graduate student at Harvard at the time, Drukier and Freese carried out the relevant calculations for photinos and neutralinos (which are predicted by the theory of supersymmetry described in chapter 10), but also for hypothetical beasts with funny names like technibaryons, cosmions, familons, and shadow matter.

In a June 1986 *Physical Review D* paper, the three theoretical physicists wrote, "If the missing mass of the Galaxy is composed of massive particles that interact with nuclei through the [weak force], SSCDs [superheated superconducting colloid detectors] can be used to detect these particles."[3] Here was a solid prediction with a clear goal, and almost immediately, experimentalists took up the challenge.

For instance, in California, Bernard Sadoulet, David Caldwell, and Blas Cabrera started the Cryogenic Dark Matter Search (CDMS), initially using germanium crystals as their target material.[4] The team's first findings were published in 1988. They didn't detect dark matter, but those null results ruled out some of the more exotic candidates and provided valuable upper limits on WIMP interactions.

CDMS really took off in the 1990s, after Sadoulet established the Center for Particle Astrophysics in Berkeley; this was the work Richard Gaitskell got involved in, in 1995. The team painstakingly built ever larger and more sensitive detectors of both germanium and silicon crystals, and they finally installed their CDMS II experiment in an abandoned, 700-meter deep iron mine in Soudan, Minnesota, in 2003.

By that time, many other groups had entered the dark matter hunt, using similar technologies. One noteworthy experiment using germanium crystals is EDELWEISS, deployed around the turn of the century in the Modane Underground Laboratory in the Fréjus

Dan Bauer, project manager of the Cryogenic Dark Matter
Search, removes germanium- and silicon-crystal detectors
from the CDMS II experiment at the Soudan Mine in
Minnesota.

Road Tunnel, which runs beneath the French–Italian border.[5] And
as we saw in the previous chapter, scientists were also experimenting
with liquid noble gases instead of cryogenic crystals.

The *Physics Review D* publication by Drukier, Freese, and Spergel
had initiated a cottage industry of dark matter search programs, and
in the second half of the 1990s, the general feeling was that the
riddle would be solved within a couple of years.

Apart from predicting WIMP interaction rates, the comprehensive 1986 paper also predicted that the dark matter wind should blow stronger and weaker at predictable points over the course of a year. "The Earth's motion around the Sun will produce a distinctive modulation in the signal detected from halo particle candidates," the authors wrote. "This modulation effect will be important in any dark-matter detector with reasonable energy resolution. . . . This signal modulation will allow additional confirmation of detection."

And that's where Rita Bernabei's DAMA experiment comes in.[6]

Bernabei, born in 1949, has been affiliated with the Tor Vergata University of Rome since 1986. Together with her younger colleague Pierluigi Belli, she started DAMA in the early 1990s, right on the heels of CDMS. But while CDMS used germanium and silicon, the first version of the Italian instrument contained nine 10-kilogram scintillation detectors of ultra-pure thallium–doped sodium iodide—known as NaI(Tl)—as its target material. Shielded against natural radioactivity from the surrounding rock by 10 centimeters of copper, 15 centimeters of lead, and one meter of concrete, the experiment, known as DAMA / NaI, was placed in the Gran Sasso laboratory, out of reach for the majority of cosmic rays.

The earliest presentation of preliminary DAMA results, in September 1997, created a stir. At the first International Conference on Particle Physics and the Early Universe in Ambleside, in the scenic Lake District in northern England, Bernabei, Belli, and their colleagues claimed a tentative detection of an annual modulation in their data.[7] Although the analysis involved only ten days of data-taking in the summer and about four weeks in the winter, it looked like DAMA was indeed seeing slightly more interaction events in June and fewer in December, just what you would expect from a genuine WIMP signal.

You might think that physicists and cosmologists would have embraced the DAMA result. After all, the mystery of dark matter had plagued scientists for decades, so every hint of a direct detection, no matter how insubstantial, would seem to be cause for joy and celebration. Instead the conference participants responded skeptically. What percentage of the scintillation flashes observed by DAMA could be credited to WIMPs? What about the statistical significance? Might there be a more mundane explanation for the observed effect? Can we have a look at your raw data and do our own analysis?

In case you're wondering about this lack of enthusiasm, or you're feeling sorry about the research community's apparent antagonism toward the DAMA results, you have to realize that this is common scientific practice. You can't believe a result just because you like it. Each and every claim should be rigorously interrogated. And without independent confirmation, nothing really counts. When they published their first results in 1998, even the DAMA team members themselves were careful enough to write, "Considering both the difficulty of this kind of searches [*sic*] and the relevance of a positive result, a cautious attitude is mandatory."

But as the DAMA / NaI experiment kept operating for months, then years, the effect wouldn't go away. Instead the modulation pattern became more evident as more data came in: a perfect sine wave, peaking around June 2 and reaching its low point around December 2—precisely what Drukier, Freese, and Spergel had predicted for a WIMP wind. When Bernabei's team published their analysis of seven years' worth of data in 2003, they were much more confident, arguing that "the WIMP presence in the galactic halo is strongly supported."[8]

DAMA / NaI was subsequently upgraded to DAMA / LIBRA (LIBRA stands for Large sodium Iodide Bulk for RAre processes). With twenty-five crystals instead of nine, and a total target mass of almost 250 kilograms, the experiment was now much more sensitive, and the evidence for a variable event rate only became sturdier. More in early June, fewer in early December, year after year.

Most physicists remained skeptical, though. If DAMA / LIBRA was really seeing dark matter, why would other instruments fail to detect anything? Granted, there were a few tantalizing hints from other experiments, but they weren't that convincing and could very well be due to background events. And over the years, it became harder and harder to reconcile the DAMA / LIBRA claims with the null results from Elena Aprile's XENON program, which was running in the same Gran Sasso lab and was getting more sensitive with every upgrade. Meanwhile Bernabei still declined to share the raw data of her experiment, and her team never published their own estimates of the frequency of background events like detections of cosmic rays.

Of course, both skeptics and the DAMA scientists themselves have attempted to explain the modulation signal without invoking dark matter. Who knows, maybe it's all due to some mundane seasonal effect, like a tiny variation in temperature or air pressure. If it's not dark matter, there has to be some other explanation. After all, it's physics—not magic.

But no matter how hard researchers have tried, no one has come up with an alternative explanation that reproduces the neat June-December sinusoidal pattern. Nothing works. As Bernabei and her colleagues noted in their 2013 paper on the DAMA / LIBRA-phase I results, "No systematic or side reaction able to mimic the exploited [modulation] signature has been found or suggested by anyone over more than a decade."[9] Five years later, the team found the same annual variation in the second phase of the experiment, which started in 2011 after yet another (relatively minor) upgrade.[10]

And that's where things stand. For almost thirty years, Rita Bernabei has been in charge of improving the instrumentation, analyzing data, publishing papers, giving conference talks, and confronting her critics. There have been hurdles and setbacks. In early

2015 her husband and collaborator, Italian physicist Silio d'Angelo, died. Bernabei herself is now well beyond retirement age. But she's not giving up.

And she still doesn't talk to journalists.

In response to my email, in which I ask her opinion about the conflicting lack of detections in other experiments, she replies rather formally. "It is not possible to make a direct model-independent comparison among the results obtained," she explains. "For that, one has to consider wide sets of astrophysical, particle physics and nuclear physics models; in addition, a lot of experimental and theoretical uncertainties exist. In some cases important differences exist in methodological procedure."

Not exactly what I would call "better transparency."

Attempts to convincingly confirm or refute the DAMA / LIBRA results have met with limited success so far. Ideally, you would have another group construct a very similar experiment at a different location, to check if the same annual modulation is observed. By building the control experiment in the southern hemisphere, you could even rule out potential seasonal effects.

It sounds easy enough, but in practice, it's hard to beat three decades of experience and technology development. In December 2010 a team led by University of Wisconsin physicist Reina Maruyama deployed two sodium iodide scintillation detectors at the geographical south pole, at a depth of 2.5 kilometers in the Antarctic ice sheet, piggybacking on the final construction phase of the IceCube Neutrino Observatory. DM-Ice, as the experiment is called, is still taking data, but it was intended as a feasibility study, and it has insufficient sensitivity to really detect anything worthwhile.

According to Maruyama, who is now at Yale, the plan is to immerse much larger and more sensitive NaI(Tl) detector modules in the ice in 2022, when the planned IceCube Upgrade takes off and new boreholes will be drilled.[11] In the meantime, her team has joined the COSINE-100 project in the Yangyang Underground Laboratory in South Korea—a DAMA clone using eight thallium-doped sodium iodide crystals with a total mass of just over 100 kilograms (hence the number). COSINE-100 began collecting data in October 2016 and published its first preliminary results in December 2018.[12] However, "several years of data will be necessary to fully confirm or refute DAMA's results," the team wrote in their *Nature* paper.

The same is true for the ANAIS experiment (Annual modulation with NaI Scintillators) in the Canfranc Underground Laboratory in the Spanish Pyrenees. Data from ANAIS's first three years of operation, published in 2021, do not show any evidence for an annual variation in the number of detected events, but the statistical significance of the preliminary result is relatively low.[13]

Yet another "control experiment" under construction is known as SABRE.[14] The plan is to run two identical copies of a DAMA-like detector: one at Gran Sasso, and one in the southern hemisphere, in the Stawell gold mine in Victoria, Australia. As of this writing, the Italian half of SABRE is still in the proof-of-principle phase, while the development of the Stawell Underground Physics Laboratory has suffered many delays. Science is not only hard, it's also slow.

Thirty-five years after Andrzej Drukier, Katherine Freese, and David Spergel published their paper on the prospects of direct detection of WIMPs in the galactic halo, the true nature of dark matter is as mysterious as ever. And although physicists have built a large number and a wide variety of sensitive experiments, in every mine, cave, and tunnel they can lay their hands on, DAMA / LIBRA is still the only one that claims to see an annual modulation in the number of particle interactions—again, a possible indication of the expected yearly variation in the speed of the WIMP wind.

But there's another important property of this supposed dark matter flow that scientists would love to measure. Just like any atmospheric air flow, the WIMP wind not only has a certain speed, it also has a definite direction. As our solar system orbits the center of the Milky Way galaxy, it is moving in the direction of the bright star Vega in the constellation Lyra. Therefore the WIMP wind is expected to preferentially blow from that particular part of the celestial sky.

Detecting the collision between an incoming particle and an atomic nucleus doesn't tell you where the intruder is coming from. If physicists could measure this direction for every interaction they register, it would be much easier to discriminate between genuine WIMPs and background events and to conclusively prove the existence of dark matter: theoretical models suggest a ten-to-one asymmetry between the number of dark matter particles coming from the direction of Vega and the flux from the opposite direction.

The best way—in fact the only way—to track down where an impacting WIMP is coming from is by measuring the direction in which the target nucleus is kicked forward. Unfortunately, most nuclei will be jolted by no more than a few nanometers before they come to rest again. That's less than one-ten-thousandth of the width of a human hair—much too small to discern. But if your target material has a very low density—if, say, it is a rarefied gas instead of a solid crystal or a dense liquid—it will take longer before the harassed nucleus comes to rest. As a result, nuclear tracks may be as long as a few micrometers and could be resolved with current detector technology.

Of course, the chance that a dark matter particle will collide with an atomic nucleus is much lower in a tenuous gas than in a solid crystal, so to detect anything at all, you may need a volume four times as large as an Olympic swimming pool. So far, no one has built such a humongous detector (mind you, it has to be deep underground!), but a small prototype called DRIFT (Directional Recoil Identification from Tracks) is operational at the Boulby mine in northern England. In the future, a much larger version of the experiment might actually be able to measure the direction in which the WIMP wind is blowing, provided the mysterious dark matter particles are detected in the first place.

If by now you are impressed by the cleverness of experimental physicists, get ready for the next flabbergasting idea: dark matter hunters have teamed up with geneticists to use human DNA as their directional detector. Don't worry, we're not talking about genetically modified organisms that glow green when hit by a WIMP (come to think of it, that sounds like another neat concept), but the biological detectors first described in 2014 are almost as ingenious.

As Freese recounts in her book *The Cosmic Cocktail,* it all started with a crazy plan by her former collaborator Drukier.[15] The Polish physicist realized that DNA can be used as the target material in a tracking chamber—a device used to track the three-dimensional path of an energetic particle. Before long Drukier and Freese were discussing the biological details with world-famous geneticists Charles Cantor and George Church, and soon enough it became clear that yes, this should work, although there were numerous hurdles to overcome.

The concept is simple enough. Thousands of identical detector modules form, basically, a hanging forest of identical DNA strands suspended from a piece of gold foil. (Imagine something that looks a little bit like a waterfall showerhead with individual water streams.)

An incoming WIMP knocks a gold nucleus out of the foil, which is only a few nanometers thick. The heavy nucleus subsequently plows across a few micrometers through the DNA forest before it loses momentum.

The key thing is that DNA is extremely vulnerable. As soon as an energetic gold nucleus slams into the long molecule, the strand is cut in two, and the lower part falls down onto a capture foil. Using the common molecular biology technique known as polymerase chain reaction (also used in COVID-19 PCR tests), scientist can quickly produce billions of identical copies of each broken-off piece of DNA, to study it in detail. And because each DNA strand has a unique sequence of bases, researchers can figure out exactly where the hanging strand was cut in two. Combining this "vertical coordinate" with the location of the DNA segment on the capture foil enables one to reconstruct the three-dimensional path of the recoiling gold nucleus with nanometer accuracy. That, in turn, gives you the arrival direction of the incoming WIMP.

It's a mind-boggling concept. Drukier, Freese, Cantor, and Church, together with Spergel, physicist Alejandro Lopez, and biologist Takeshi Sano announced their ideas online in the summer of 2012 and expanded on them in a 2014 paper in the *International Journal of Modern Physics A*.[16] However, an operational DNA-based dark matter detector is still a long way off.

So far, no one has succeeded in measuring which way the WIMP wind blows. Worse, there's no consensus on whether or not physicists have actually detected the expected annual modulation in the wind speed. In that respect, the DAMA experiment remains the odd one out. The answer is blowing, well, you know the Bob Dylan song.

Talking about Bob Dylan: maybe we should also keep knockin' on heaven's door. After all, our home planet—let alone a physics experiment in an underground laboratory—constitutes a pretty tiny search area for the elusive dark matter particles we're after. There's much more space out there, and much more mysterious stuff than what we may expect to encounter here on Earth. Might it be that the huge amounts of dark matter particles suspected of lurking in the center of our galaxy betray their existence through some extremely rare interaction? Could we use the whole Milky Way as our particle detector?

Once more, it's time to train our gaze on the heavens.

20

Messengers from Outer Space

On February 3, 2020, a forty-three-year-old man posts a selfie on Twitter. "At work, checking the [sewage] for leaks," he writes.[1] Not something to pay much attention to, you'd say. Except that this man is Italian astronaut Luca Parmitano. The tweet is sent from the International Space Station. The photo shows Parmitano during his fourth spacewalk in ten weeks. And the sewage he is checking, a set of cooling pipes, is part of a $2 billion-dollar experiment to search for antiparticles and dark matter.

Thanks to repairs made by Parmitano and NASA astronaut Andrew Morgan, the Alpha Magnetic Spectrometer (AMS), which has been mounted on the outside of the space station since May 2011, got a new lease on life. Much to the delight of principal investigator and Nobel Laureate Samuel Ting, who has been in charge of the project for twenty-five years. During a Zoom interview from his house on the US East Coast, Ting tells me he is confident that AMS will provide the definitive answer on the nature of dark matter by 2028.[2]

The dark matter detectors described in the previous two chapters are located in deep underground laboratories, to shield them from cosmic ray particles. In outer space, cosmic rays are everywhere, so there's no chance whatsoever that AMS will directly detect dark matter in the same way crystal and xenon experiments are trying to. Instead the hope is to find antimatter particles that may result from dark matter annihilation. So far the results have been surprising and exciting, to say the least.

Physicists have known about the existence of antimatter since 1932. For every type of elementary particle, there is a corresponding antiparticle with exactly the same mass but with opposite electrical charge and magnetic moment. However, dark matter particles have zero charge, and according to most models they also have zero magnetic moment. If so, they can be their own antiparticles, and when two of them collide, they may annihilate each other. As $E = mc^2$ tells us, the total energy of the annihilation would give rise to a cascade of ordinary particle / antiparticle pairs, including protons (hydrogen nuclei) and antiprotons, as well as electrons and positrons (anti-electrons).

Physicists are used to the presence of particle / antiparticle pairs in space. For instance, they are produced by nuclei from supernova explosions colliding at high velocity with atoms in interstellar space. But if high-energy antiprotons and positrons appear in excess of expected numbers, that might point to annihilating dark matter. This possibility is what got physicists searching eagerly for antimatter in cosmic rays—the continuous downpour of all types of energetic particles from outer space. You cannot do that from the ground, however. Cosmic rays produce showers of secondary particles upon entering Earth's atmosphere, and terrestrial detectors can't tell you much about the nature of the primary particles. If you want to hunt for cosmic antimatter, you need to have a magnet and you need to go into space.

Enter particle physicist Samuel C. C. Ting. Born in the United States, Ting spent his childhood in China and Taiwan. He returned to the United States in 1956 at age twenty and has been at the Massachusetts Institute of Technology since 1969. In 1976 Ting shared the Nobel Prize in Physics with Burton Richter of the Stanford Linear Accelerator Center for their independent 1974 discoveries of the surprisingly long-lived J/ψ meson, the first particle found to contain a charm quark, as well as a charm antiquark. Frustrated by the 1993 congressional decision to cancel the planned Super-conducting Super Collider—a huge underground accelerator that would have secured the future of American particle physics—Ting decided to change gears and set his sights on space and the search for antimatter.

As luck would have it, 1993 was also the year in which NASA, together with other space agencies around the globe, agreed to co-operate on what would eventually become the International Space Station (ISS). Fundamental science was touted as one of the main justifications for the enormous amount of taxpayer money that would go into building, launching, assembling, and operating the ISS, so it seemed only appropriate to use the future orbiting labora-tory as the platform for the precise, massive, and power-hungry magnetic spectrometer that Ting had in mind. This would be space station science at its best.

Just like any terrestrial particle detector, the Alpha Magnetic Spec-trometer would employ a range of technologies to measure the mass, charge, velocity, and energy of nearly every cosmic ray particle that passed through its innards. A giant, 1,200-kilogram magnet at the core of the experiment, 3,000 times stronger than Earth's mag-netic field, would bend the paths of electrically charged particles, enabling scientists to discriminate between positively and negatively charged ones. And since everything was so novel, NASA required a technology-qualification flight on the space shuttle before the in-strument's long-duration mission on the space station.

Most people wouldn't know where to start, but Sam Ting didn't know how to stop, once he had the plan in his head. He won the support and cooperation of NASA administrator Daniel Goldin, secured substantial seed funding from the Department of Energy, and assembled an international scientific collaboration with members in sixteen countries. The AMS project was approved in April 1995. The first version of the antimatter detector, with components from various institutes in Europe and Asia, was completed at CERN in less than three years.

On June 2, 1998, Ting witnessed the launch of AMS-01 onboard the space shuttle *Discovery,* which was making its final flight. The shuttle was headed to the Russian space station Mir, but AMS-01 operated from the shuttle's payload bay. The instrument worked just fine, even though the shuttle mission was plagued by data-transmission problems. The experiment even yielded some unexpected scientific results, including a low-energy excess of positrons. The findings were published in 2000 in *Physics Letters B.*[3]

Originally the plan was to fly the same instrument—or at least a very similar copy—on the ISS. But it would be years before the ISS was completed, so Ting's team had ample time to improve on the design. Eventually AMS-02 measured 5 by 4 by 3 meters (it would probably not fit in your living room) and weighed 7.5 tons, almost twice as heavy as its predecessor.[4] The various detectors of the giant instrument contained 50,000 optical fibers and 11,000 photo sensors. With its 300,000 electronic channels and 650 fast microprocessors, AMS-02 produced 7 gigabits of data per second, consuming 2.5 kilowatts of power in the process. Launch was planned for 2005.

Then, on February 1, 2003, disaster struck, when the space shuttle *Columbia* disintegrated during its atmospheric reentry, killing the seven-person crew. The next year the George W. Bush administration set a new course for the shuttle program: all future flights would be devoted to ISS logistics and ferrying up spare parts. Immediately after completion of the cosmic outpost, the space shuttle fleet would be retired. There would be no room for a dedicated AMS flight anymore.

Ting was devastated. AMS was designed to fit in the space shuttle's payload bay. Without the shuttle, it couldn't fly. He tried to persuade NASA's new administrator, Michael Griffin, to add one more flight to the space shuttle launch schedule, but Griffin was legally obliged to follow the White House's orders.

The decision to cancel AMS was published across the major media. "Only after members of the US Congress became aware of the absence of major scientific research onboard the ISS did the scientific potential of AMS, and the US tradition of honoring international agreements, earn bipartisan support," Ting recounts. Thanks to his tenacity, NASA was eventually ordered to add one additional flight to the space shuttle manifest, and the future of the Alpha Magnetic Spectrometer was secured. Ting's dream would come true after all.

AMS-02 was launched on May 16, 2011, on the twenty-fifth and final mission of space shuttle *Endeavour,* to the International Space Station. Three days later the shuttle crew used the robotic arms of both *Endeavour* and the ISS to remotely install the giant instrument on the third segment of the space station's central truss. Data-taking commenced almost immediately and has continued ever since.

By the time it was launched into space, AMS was no longer the only player in the field. An Italian-led European collaboration had also built an antimatter detector, called PAMELA, for Payload for Antimatter-Matter Exploration and Light-nuclei Astrophysics. About the size of a large barrel, and weighing less than 500 kilograms, PAMELA was much smaller—and less sensitive—than AMS, but it had been taking data since June 2006, when it was launched as a piggyback instrument on the Russian Earth observation satellite Resurs-DK1. Would David beat Goliath in the search for dark matter messengers from outer space?

In August 2008, at conferences in Philadelphia and Stockholm, the PAMELA collaboration presented their first preliminary results.

The Alpha Magnetic Spectrometer (AMS), mounted on the truss of the International Space Station.

(The full analysis of the first two years of PAMELA data was published in April 2009 in *Nature*.[5]) Since PAMELA had been operating for much longer than the shuttle-borne AMS-01 did, it had captured enough rare high-energy particles to conclude that the positron excess reported in 2000 by Ting's team was still present at energies above 10 giga-electronvolt (that's a billion electronvolt, or GeV).

The new results generated a lot of buzz. What could be the source of all these anti-electrons from outer space? Well, dark matter annihilation! "If it's true, it's a major discovery," University of Chicago particle physicist Dan Hooper told *Nature* reporter Geoff Brumfiel.[6] Then again, it wasn't easy to exclude possible alternative mechanisms. Antimatter could also be produced in the high-energy environments of pulsars—rapidly rotating and highly magnetized neutron stars.

There was yet another reason why 2008 saw so much excitement about the prospect of an indirect detection of dark matter in outer space. On June 11 NASA launched its Fermi Gamma-ray Space Telescope, named after Italian American physicist Enrico Fermi.[7] With its huge field of view covering almost 20 percent of the sky, Fermi's Large Area Telescope might be able to detect the expected gamma ray glow from dark matter in the center of our Milky Way galaxy.

Of course dark matter is dark, and not supposed to emit any kind of electromagnetic radiation. But annihilating dark matter does produce high-energy photons, either directly (again, $E = mc^2$), or as part of the convoluted decay chain that ultimately yields particle / antiparticle pairs. Since the density of dark matter in the Milky Way is expected to be highest in the galactic center, you would expect a much stronger annihilation signal from that direction.

One of the scientists eager to learn about the Fermi results was Douglas Finkbeiner of the Harvard-Smithsonian Center for Astrophysics. Tracy Slatyer, who is now an assistant professor of physics at MIT, was one of Finkbeiner's PhD students at the time. She recalls how impatient he became when the Fermi team announced that, on August 25, 2009, they would publicly release their first year of data. "Doug had figured out the URL of the web page on which the full data set would be made available," Slatyer says. "He was continuously refreshing the page, to make sure we would be able to start our analysis as soon as possible."[8] As a result, the Harvard team was already downloading the data set before potential competitors were aware it was available at all.

Working with Finkbeiner, postdoc Greg Dobler, and New York University physicists Ilias Cholis and Neal Weiner, Slatyer indeed found an excess of gamma rays from the galactic center, which they called the Fermi haze. And although their discovery paper, published on the arXiv preprint server on October 26, only mentioned the possibility very briefly, dark matter was always at the back of Slatyer's mind. True, the gamma ray excess could also be due to a process known as inverse Compton scattering, where photons are kicked up to very high energies by relativistic particles. But even then, those particles had to originate somewhere—they might as well be electrons and positrons produced by dark matter annihilation.

The Fermi haze paper was finally published in *The Astrophysical Journal* in July 2010.[9] But by then the story had become even more exciting. Originally the Fermi haze had looked a bit egg-shaped, but in early 2010, when Finkbeiner's team carried out a more sophisticated analysis on a larger volume of satellite data, they realized that the shape of the gamma ray excess was more like a number 8—a vertical lemniscate. It looked like the galactic center was spewing out two huge bubbles into space, in opposite directions along its axis of rotation.

Finkbeiner, Slatyer, and fellow PhD student Meng Su announced their discovery of the "Fermi bubbles" in a paper posted on the arXiv preprint server on May 29. When Harvard issued a press release on November 9, accompanied by an impressive artist's rendering of the giant bubbles, the story hit newspapers all over the world. The paper was eventually published in the December 1 issue of *The Astrophysical Journal*.[10]

What the heck was going on in the core of the Milky Way? What mechanism could possibly produce two giant bubbles of gamma ray emission, each some 25,000 light-years across, above and below the central plane of our galaxy? Additional observations strongly suggested that the bubbles are the result of a fast outflow, probably due to some explosive event at the galactic center, maybe just a few million years ago. Energetic electrons and other charged particles in the outflow would produce gamma rays through the aforementioned process of inverse Compton scattering.

Initially Slatyer was a bit disappointed to learn that the bubbles were probably not the key to unlock a major mystery. "It would've been spectacularly interesting if the Fermi bubbles could be explained by dark matter," she says. Then again, if you're after an elusive—possibly nonexistent—dark matter annihilation signal in the Fermi data, you really need to know about all other mechanisms that produce gamma rays. "The bubbles must be understood in order to use measurements of the diffuse gamma-ray emission in the inner Galaxy as a probe of dark matter physics," the three researchers wrote.

At the University of Chicago, Dan Hooper couldn't agree more.[11] His independent analysis of the Fermi data revealed an additional excess of relatively low-energy gamma rays (just a few GeV) in a much smaller, roughly spherical area around the galactic center. This further concentration of gamma rays was apparently unrelated to the bubbles and possibly resulted from a different production mechanism. After convincing Slatyer of the validity of his result, the two researchers joined forces, and in 2013 they published a detailed paper in the online-only journal *Physics of the Dark Universe*.[12]

Might this excess of low-energy gamma rays be due to annihilating dark matter after all? Or was that just wishful thinking? A large population of millisecond pulsars—pulsars that spin very rapidly, completing hundreds of rotations every second—could be responsible too, but how would these exotic objects end up so far above and below the central plane of the Milky Way, up to ten-thousand light-years away from the galactic center?

The issue is still unresolved. There have been many strong arguments against the dark matter interpretation, but, more recently, also strong arguments against those arguments. Around 2017, millisecond pulsars were generally seen as the leading candidate explanation. But two years later, in a paper in *Physical Review Letters,* Slatyer and MIT colleague Rebecca Leane concluded that "dark matter may provide a dominant contribution to the galactic center excess after all."[13]

Pulsars or dark matter? It's the same question that keeps the theorists of the international Alpha Magnetic Spectrometer collaboration awake, says Mercedes Paniccia, an astroparticle physicist at the University of Geneva, who helped build the silicon trackers for AMS-02. When I planned a visit to CERN in June 2019, I had hoped to meet AMS principal investigator Samuel Ting there; he usually spends most of his time at the European particle physics lab. But apparently he was going to be away at some conference. "He's hard to catch," Paniccia told me when I got in touch with her, "but I can show you around. Just meet me at the AMS POCC, that's building 946."

The Payload Operations Control Center, a few kilometers away from the main CERN gate, holds a permanently staffed, high-tech scientific nerve center.[14] Physicists and technicians are working at computer consoles that line two sides of the room. Giant wall screens provide live views from the ISS and from Mission Control at NASA's Johnson Space Center in Houston. A huge digital world map keeps track of the space station's actual position. And I can't take my eyes off the large display beneath the map that shows the number of cosmic rays detected since AMS-02 started operations. During my half-hour stay, the glowing red numbers increase from 139,767,027,021, with the ISS flying above Africa, to 139,768,372,421, as the space station is crossing the Pacific Ocean. "That's about 600 events per second," Mercedes says. "Most of them are protons. Electrons come second, but we also detect numerous heavier atomic nuclei, antiprotons, and positrons." Some of which may result from annihilating dark matter, I can't help thinking, looking at the digital counter again.

At the head of the central conference table is a large nameplate reading, "Prof. Samuel C. C. Ting" and an enormous, empty leather desk chair. "I'm not sure of his whereabouts," Mercedes says. "He's extremely busy."

Back home, I try to get in touch with Ting through at least three different email addresses, but there's no reply. "It will probably be hard to make an interview appointment," Amsterdam dark matter researcher Suzan Başeğmez tells me. "He's a Nobel laureate, you know." Ting's personal assistant at MIT, Christine Titus, also has bad news for me: during my East Coast trip in January 2020 he will be traveling. I almost give up.

Finally, in September 2020, after yet another email request, Ting gets back to me. Yes, we can do a Zoom interview later that month. It's the day of my fortieth wedding anniversary, but I don't mind. I expect a brief talk of fifteen minutes or so, and I prepare a short list of questions. Instead the eighty-four-year-old physicist treats me to a ninety-minute private lecture, supported by many dozens of Powerpoint slides. In his characteristic soft voice, he talks about particle physics, antimatter, detector technology, and politics. And he shares personal anecdotes, too, like how he nearly lost all faith in his project after the AMS-02 flight had been cancelled in response to the *Columbia* accident. Or how he got a $245 fine for speeding when he raced to the Kennedy Space Center to witness the particle detector's launch. Or how worried he was, just before *Endeavour* took off, that something might go wrong—that all those years of effort had been in vain.

Twenty-five years after he first came up with the idea of a large particle detector in space, Samuel Ting has now captured more than 150 billion cosmic ray particles, including a few million positrons, but he is still reluctant to claim the discovery of annihilating dark matter. Asked whether he thinks that dark matter may explain the AMS data, he replies, "What I think is not meaningful. The data provide a strong hint, but not yet a proof."

Yes, there are many more positrons at energies between 3 and 1,000 GeV than can be accounted for by known astrophysical processes, and the energy distribution doesn't fully agree with the data from the much smaller PAMELA experiment, which ceased operations in 2016. But the AMS results didn't provide definitive proof of decaying dark matter: in principle, the observed excess (first described in 2013 in *Physical Review Letters*) could be due to a relatively small number of high-energy pulsars in our galactic neighborhood—unfortunately, charged particles cannot easily be traced back to their points of origin.[15]

Then again, AMS-02 also discovered a similar excess of antiprotons, and pulsars are not energetic enough to produce these much more massive antiparticles. If positrons and antiprotons are produced by two different mechanisms, it would seem a bit unlikely—if not a conspiracy of nature—that they would exhibit the same energy spectrum. Instead, if both types of antimatter particles are the result of dark matter annihilation, you would expect them to show more or less similar behavior, which is what the AMS-02 data show.

Pulsars or dark matter, again. A longer operational lifetime for AMS-02, much more cosmic ray data, and a correspondingly larger number of very-high-energy antimatter particles may eventually settle the question. That's why Ting is so happy with the four successful six-hour spacewalks that astronauts Luca Parmitano and Drew Morgan carried out between mid-November 2019 and late January 2020. Three of the four pumps of the AMS cooling system had broken down over the years, and all four were replaced by new, heavy-duty ones, in a series of challenging and time-consuming space repairs, closely watched by Ting from Houston mission control. And yes, on the fourth and final spacewalk, Parmitano fixed a leak in one of the cooling pipes, or sewage, as he jokingly called it in his February 3 tweet.

The International Space Station will remain operational until at least 2028. By then, says Ting, AMS-02 will have detected so many high-energy antimatter particles that it will become possible to compare the data with the predicted energy spectrum for annihilating dark matter in the galactic center. Meanwhile, scientists are already discussing a much larger and more sensitive space-based cosmic ray detector, provisionally called AMS-100, that could be launched around 2040. If realized, I predict that it will be named after Sam Ting.

Tracy Slatyer and Dan Hooper both believe that the mystery of the gamma ray excess from the Milky Way's core may be settled within a few years. In particular, they have high expectations for large radio observatories like the MeerKAT array in South Africa and the future Square Kilometre Array (SKA). "If there really is a large population of energetic millisecond pulsars in the galactic center," says Slatyer, "a deep radio survey with SKA should be able to find hundreds of them." According to Hooper, pulsar experts can only account for the gamma ray observations by assuming the existence of a whopping 3 million individual sources. "If SKA doesn't find any, we can safely reject that explanation," he says.

But that would not necessarily reveal the origin of the excess of high-energy positrons and antiprotons observed by AMS-02—according to Slatyer, they could still be produced in nearby pulsars. "It's unlikely that all of these excesses arise from the same dark matter source," she says.

The messengers from outer space keep raining down on our tiny planet, both in the form of cosmic ray particles and as high-energy gamma photons. Somewhere in this interstellar avalanche, scientists may locate the key to the indirect detection of dark matter. For now, the haystack is still too large to find the needle.

21

Delinquent Dwarfs

In the same year Sam Ting finally witnessed the launch of his $2 billion antimatter detector, Pieter van Dokkum and Roberto Abraham first discussed plans for a low-budget, ultrasensitive camera system to hunt for ghost galaxies and other faint structures in the night sky.[1]

As chair of the astronomy department at Yale, van Dokkum sometimes felt lost in big projects, grant applications, and organizational meetings. Abraham, at the University of Toronto, felt the same way. In 2011, during a dinner in downtown Toronto, the two friends reminisced about the good old days, when doing science was just fun. Where did it all go wrong? And what about bringing back the youthful excitement of the past by starting a new, hobby-like project?

Two years later, the first small version of the Dragonfly Telephoto Array was commissioned at the New Mexico Skies Observatories. "I don't remember whose idea it was originally," says Abraham. "We'd had a few beers. I think it was both of us."

You'll read more about Dragonfly later in this chapter—it's a really exciting project.[2] For now, let's just fast-forward to late March 2018, when Dragonfly hit the headlines by apparently providing proof for the existence of dark matter, while debunking the alternative-gravity theory of modified Newtonian dynamics (MOND) discussed in chapter 12. Not by detecting the mysterious stuff but by turning up a dwarf galaxy that appears to be completely devoid of it.

Van Dokkum calls it a Zen-like argument: proving the existence of something by not finding it. It all makes sense if you think about it. If the high rotational velocities of galaxies are caused by some unrecognized behavior of gravity, as claimed by MOND, each and every galaxy should show the same effect. But if the rotational velocities are due to dark matter, as most astrophysicists assume, galaxies without dark matter should rotate more slowly, at a pace you would expect on the basis of the observed quantity of stars and gas alone. Dwarf galaxy NGC 1052-DF2 did just that. It couldn't be governed by MOND.[3] Ergo dark matter must exist. (Or at least MOND is wrong.)

No one could easily explain the existence of a galaxy containing no dark matter at all, but the DF2 discovery underlined the importance of dwarf galaxies in the study of dark matter.

Astronomers have known about two small companion galaxies to the Milky Way for ages—the Large and the Small Magellanic Clouds, easily visible with the naked eye from the southern hemisphere and the Tropics. Andromeda also has two prominent satellite galaxies, first observed in the eighteenth century by French astronomer Charles Messier. Still, the discovery of the first "real" dwarf galaxies—much smaller than the Magellanic Clouds—by Harvard astronomer Harlow Shapley in 1937, came as a surprise. In a letter to the editor in the October 15, 1938, issue of *Nature,* Shapley speculated that "such objects may be of frequent occurrence in intergalactic space."[4]

Indeed, as of this writing, our Milky Way galaxy is known to host at least fifty-nine dwarf satellites within a distance of 1.4 million light-years. The larger ones come in a wide variety of shapes and types, from amorphous, gas-rich aggregates of star clusters and nebulae to highly symmetrical systems of old stars that look like miniature versions of elliptical galaxies. On average, these satellites are tens of times smaller than their host galaxy, and they usually contain a few hundred million stars at most, as compared to the Milky Way's 400 billion or so.

So what do dwarf galaxies tell us about dark matter?

Well, unbeknownst to Shapley back in the 1930s, they are believed to mark the dark matter building blocks of the cosmos. Remember that the most successful supercomputer simulations of structure formation in the universe present us with a bottom-up scenario of hierarchical clustering. As we saw in chapter 11, cold dark matter particles have a relatively low velocity, so they first gravitationally clump into small dark matter halos. These dark matter clumps will start to accrete normal, baryonic matter, from which new stars are born. Over time, a fair fraction of the resulting dwarf galaxies are expected to merge into full-grown systems like our Milky Way.

Although the cold dark matter theory doesn't allow you to make firm predictions from first principles, the simulations show a consistent picture. Large galaxies—each embedded in its own almost spherical dark matter halo—are surrounded by countless so-called subhalos: basically, dark matter–laden dwarf galaxies that have not yet been consumed by the central heavyweight and maybe never will.

By studying the outcomes of large supercomputer simulations like IllustrisTNG and EAGLE in much detail, astronomers can "predict" the properties of dwarf galaxies. Conversely, observing and studying real dwarf galaxies is a good way to check the validity of the now-popular ΛCDM model—the cosmological concordance model with cold dark matter and dark energy, on which the simulations are based.

The good news is: these checks have been made. The bad news is that dwarfs are misbehavers. They don't act like they're supposed to do.

To begin with, they're not nearly as numerous as they should be. Sixty satellite dwarfs swarming around the Milky Way may sound like a lot, but theorists predict there should be at least five hundred. And it's not that astronomers haven't searched hard enough. The current surveys should really have turned up many, many more. It's called the missing-satellite problem, and it's real.

Scientists have proposed many potential solutions to the missing-satellite problem. For instance, we know that, over the past 10 billion years or so, our Milky Way galaxy has devoured dwarf galaxies that ventured too close, slowly ripping them apart through tidal forces and eventually gobbling up both their constituent stars and their dark matter inventory. So maybe the feast is almost over, and that is why the Milky Way is now left with only a few dozen satellites.

Another possibility is that there really are many hundreds of dark matter subhalos out there, but for some reason they were unable to produce significant numbers of new stars, rendering them invisible to our telescopes. Imagine: huge blobs of mysterious dark matter and hardly anything else, slowly orbiting our Milky Way galaxy in all possible directions.

However, it has been hard, if not impossible, to solve the missing-satellite problem in a satisfying way. The numbers just don't match. Given the masses and luminosities of the observed satellites, ΛCDM predicts that there should be quite a few larger and more massive subhalos that definitely should have turned into conspicuous dwarf galaxies—a discrepancy known as the too-big-to-fail problem.

Dwarfs misbehave in other ways, too, further complicating the dark matter search. Consider the effort to estimate the dark matter contents of dwarf galaxies by studying the rotational velocities of the stars and gas clouds within them. Such measurements were pioneered in the 1980s by University of Arizona astronomer Marc Aaronson, who was killed in 1987 by a freak accident in the dome of the 4-meter Mayall Telescope at Kitt Peak National Observatory. Around the turn of the century, the project was carried out in more detail by John Kormendy and Ken Freeman, who collected and analyzed data, obtained by others, on a wide variety of galaxies.

What Kormendy and Freeman found was that dwarf galaxies contain a higher fraction of dark matter than big spiral galaxies do, and dwarf galaxies are also more densely laden with the stuff.[5] So far, so good: this is in good agreement with the ΛCDM computer simulations. But these same simulations also produce subhalos with a very characteristic density profile: as you move toward the core, the dark matter density increases faster and faster, until it reaches a peak value at the very center. This density profile appears to be an inescapable result of hierarchical clustering, at least in the supercomputer simulations.[6]

The problem is that real dwarf galaxies do not show these prominent density cusps at their cores. The dark matter distribution, as derived from velocity observations, is always much flatter. This third mismatch between ΛCDM simulations and the real universe is called the core-cusp problem, or the cuspy halo problem. Again, there's no easy explanation. Dwarf galaxies just don't obey the rules. Or, put differently, our rules do not properly describe reality.

The Dragonfly Telephoto Array shed surprising new light on the relation between dwarf galaxies and dark matter. However, dark matter was not on Pieter van Dokkum's and Bob Abraham's minds in 2011, when they first discussed the possibility of using off-the-shelf photographic lenses to image extremely diffuse structures in the night sky—faint wisps of nebulosity, but also low-surface-brightness galaxies. As an avid nature photographer, van Dokkum had heard about a new professional 300-mm telephoto lens produced by Canon, featuring a nanotechnology-based coating to reduce light scattering. That sounded perfect for wildlife and sports photographers working with backlight. It also sounded perfect for low-contrast deep-sky photography.

"Off-the-shelf" doesn't necessarily mean cheap. The lens sold for approximately $10,000. But hooking up several of these lenses would still be much cheaper than designing and building a special-purpose telescope. By aiming many lenses at the same part of the sky, each mounted on its own professional CCD camera, you could digitally add up the individual images to further improve contrast and sensitivity. Thus the idea for an astronomical telephoto array was born.

It didn't take long to come up with an appropriate name for the project. A big array would look more or less similar to the compound eye of a dragonfly, as van Dokkum knew well enough: ever since he was a young boy in the Netherlands, he had taken thousands of macro images of the beautiful insects, and he was working on a book.[7] Dragonfly it would be.

What started out with test images in van Dokkum's New Haven basement and backyard, using just one lens, soon grew into a working prototype of three lenses on one mount. Hardly more than a year after the Toronto dinner, he and Abraham moved their equipment from the Mont-Mégantic Dark Sky Reserve in southern Quebec to the even darker New Mexico Skies Observatories, in the Lincoln National Forest east of Alamogordo, where semiprofessional astronomers from all over the country operate dozens of remotely controlled telescopes.[8]

Meanwhile van Dokkum and Abraham recruited students to join the Dragonfly project and to build hardware, develop software, process the data, and analyze the results. The array quickly grew from

Part of the Dragonfly Telephoto Array at New Mexico Skies Observatory.

three lenses to eight, then ten, then twenty-four on one telescope mount—an impressive compound eye indeed. Before long, the team constructed a second dome with a second twenty-four-lens array. The forty-eight telephoto lenses have the same overall collecting area as a virtual 1-meter telescope, but the focal length is still a mere 40 centimeters, yielding an incredibly "fast" optical system with a focal ratio of 0.4—a system able to register small quantities of light in a short span of time—something that can never be achieved with a single lens or mirror.

Dragonfly may have started out as a hobby project, but it soon evolved into a novel, high-profile robotic observatory that uniquely focused on the much-neglected low-contrast and low-surface-brightness universe. Not surprising, then, that it started to yield exciting scientific results right away. Dragonfly images of the Coma Cluster of galaxies revealed forty-seven extremely faint smudges of light, the vast majority of which had never been observed before. In their 2015 discovery paper in *The Astrophysical Journal Letters,* van Dokkum, Abraham, and their colleagues called the smudges ultra-diffuse galaxies, or UDGs.[9] If they were really at the same distance as the Coma Cluster (some 320 million light-years)—that is, if the UDGs were in fact part of the cluster—then they were about as large as normal galaxies but also a few hundred times fainter. This implies that they contain at most 1 percent of the expected number of stars for a galaxy of their size.

Follow-up spectroscopic observations of one of the Coma UDGs (Dragonfly 44) with the 10-meter Keck Telescope at Mauna Kea, Hawaiʻi, confirmed that the galaxy indeed belongs to the cluster.[10] Images obtained with the 8-meter Gemini North Telescope, also at Mauna Kea, further revealed that DF44 is surrounded by many dozens of globular star clusters, just like our own Milky Way, and subsequent velocity measurements yielded a surprisingly large mass, comparable to the mass of our home galaxy. Still, DF44 contains almost no stars. "Dragonfly 44 can be viewed as a failed Milky Way," the authors concluded in their second follow-up paper, published in 2016.[11] With a dark matter fraction of a whopping 98 percent, DF44 apparently represents a new class of "dark galaxies" that has never been recognized before.

The discovery of ultra-diffuse galaxies like DF44 has kept scientists busy over the past few years. No one has a good explanation for the origin of these low-luminosity monsters—they don't pop up in computer simulations of a ΛCDM universe. Neither does modified Newtonian dynamics offer a satisfactory answer: there are just not enough stars to explain the velocity measurements, even if you use MOND's alternative gravity equations. As MOND champion Stacy McGaugh says, "DF44 is a problem for everybody."[12]

Little wonder that some astronomers question the distance determination, velocity measurements, mass estimate, and even the number of globulars found by the Dragonfly scientists.

Finding "failed" galaxies that are completely dominated by dark matter caused quite a stir, but Dragonfly's next major discovery was even more controversial. As the team reported in a 2018 *Nature* publication, they had also hit upon a galaxy that appears to contain hardly any dark matter at all—a faint smudge of light close to the massive elliptical galaxy NGC 1052, which is some 65 million light-years away, much closer than the Coma Cluster.[13] From the observed amount of light, van Dokkum and his colleagues derived a stellar (baryonic) mass of some 200 million solar masses, quite typical for a relatively large dwarf galaxy. And like the Coma UDGs, this faint dwarf is surrounded by lots of bright globular star clusters.

Spectroscopic measurements with the Keck Telescope revealed the orbital velocities of ten of those globulars and enabled the astronomers to "weigh" the galaxy, just as they had done with other ultra-diffuse galaxies. However, instead of finding evidence for huge amounts of invisible dark matter, they found a total mass that is hardly larger than the baryonic mass mentioned above. In other words, NGC 1052-DF2, as the enigmatic galaxy is now known, appears to be almost devoid of dark matter.

In 2019 the Dragonfly team announced the discovery of NGC 1052-DF4, a second galaxy in the same group with very similar properties. "The origin of these large, faint galaxies with an excess of luminous globular clusters and an apparent lack of dark matter is, at present, not understood," they wrote in their *Astrophysical Journal Letters* paper.[14]

So first we had an unexplainable deficit of small dwarf galaxies, and dark matter density profiles that are much too flat. Next Dragonfly has given us weird dark matter–dominated galaxies that look like our Milky Way except they contain just 1 percent of the expected number of stars. And now we're presented with even weirder galaxies that are equally diffuse but appear to be devoid of dark matter altogether. All in all, that's a lot of inconvenient mysteries, none of which can easily be accounted for by the ΛCDM concordance model of cosmology.

And that's not all. There's yet another way in which diminutive galaxies conflict with theoretical predictions and expectations. This final puzzle has nothing to do with the physical or dynamical properties of dwarf galaxies but rather with their three-dimensional distribution in space. Simply put, they're not where they're supposed to be.

Detailed supercomputer simulations of the growth of cosmic structure, like IllustrisTNG and EAGLE, show how large galaxies like our Milky Way end up being surrounded on all sides by huge numbers of dark matter subhalos, which become visible as dwarf galaxies. In the real universe, however, the dwarf companions are not only too few in number; they also do not surround their host galaxy equally in every direction. Instead, the majority of satellite galaxies are found in a flattened disk, which does not coincide with the central plane of the host. No matter how computational astrophysicists tweak their code, they are not able to reproduce this distribution in their simulations. It's known as the planes-of-satellite-galaxies problem.

Back in 1976, when astronomers knew of only eight companions to our Milky Way galaxy (including the Magellanic Clouds), British astrophysicist Donald Lynden-Bell already noticed that most of them approximately line up in one single plane, more or less at right angles with the central plane of the Milky Way. But it wasn't until 2005 that three European astronomers—Pavel Kroupa, Christian Theis, and Christian Boily—studied the problem in much more detail, by comparing the observed distribution with cold dark matter simulations.[15] Their conclusion: the chance of ending up with such a disk-like dwarf galaxy distribution is only 0.5 percent.

Soon enough, it turned out that the planar distribution of Milky Way satellite galaxies is not unique. In 2013 a group led by Rodrigo Ibata of the Observatoire de Strasbourg, France, announced the discovery of a very similar structure around the Andromeda galaxy: about half of Andromeda's dwarf companions are located in a thin plane some 1.3 million light-years in diameter but only 45,000 light-years thick.[16] Moreover, as Ibata's team showed in their *Nature* paper, these satellites orbit their host galaxy in the same direction, indicating some common origin or dynamical evolution. And five years later, Swiss astronomer Oliver Müller and his colleagues found that many of the dwarf satellites of the elliptical galaxy Centaurus A, at a distance of 12.5 million light-years, also orbit their massive host in a thin, corotating plane.[17]

According to Müller's coauthor Marcel Pawlowski of the Leibniz-Institut für Astrophysik Potsdam, Germany, the planes-of-satellite-galaxies problem has no easy solution.[18] "It's also easy to ignore," he says, "but by now, many people are at least aware of it." In 2018, when he was at the University of California Irvine, Pawlowski wrote a comprehensive review article about the problem for *Modern Physics Letters A,* in which he discussed a number of possible solutions and argued for "why they all are currently unable to satisfactorily resolve the issue."[19]

One thing is for sure: if astronomers want to understand the nature of dark matter and its role in cosmic evolution, they have to dig deeper than ever before. Fritz Zwicky focused on the dynamics of massive clusters of galaxies—the largest gravitationally bound structures in the universe. Vera Rubin and Albert Bosma pioneered the study of the rotation of luminous galaxies like our own Milky

Way. At these large scales, the presence and influence of mysterious, invisible stuff is very evident. But any viable dark matter theory also needs to fully explain the observed properties and behavior of the unimposing denizens of deep space: the dim satellite dwarfs that swarm around their majestic host galaxies, and the ultra-diffuse "failed" galaxies hiding in the cosmic darkness.

Under the pitch-black New Mexico skies, the sensitive compound eyes of the Dragonfly Telephoto Array may well yield new surprises in the years to come. Van Dokkum and Abraham are expanding their array with more stations, hoping to arrive at a whopping total of 168 lenses. "There's no reason to stop at forty-eight," explains van Dokkum. "In principle, you could even build the equivalent of a 10- or 20-meter telescope, at a much lower cost."

Will cosmologists ever find a way to reconcile their cherished ideas about dark energy and cold dark matter with the observed properties of dwarf galaxies? No one knows, but future observations may provide a solution. Disturbingly enough though, the delinquent dwarfs are not alone in casting shadows on the popular ΛCDM model. Cosmology is facing an even bigger crisis.

22

Cosmological Tension

In the 1980s, when people in East and West Berlin still lived in two very different universes, politically speaking, Checkpoint Charlie was an intimidating and heavily guarded crossing point between communist oppression and liberal democracy. Today it is one of the most popular tourist attractions in the capital of united Germany. But less than three decades after the Berlin Wall was opened in 1989, another insuperable barrier, this time scientific in nature, manifested itself just 600 meters commie-ward of Checkpoint Charlie, in the Auditorium Friedrichstrasse. On a drizzly Saturday in November 2018, this unadorned Soviet-style building served as the intellectual battleground for a cosmological Cold War.

Some 130 scientists had flocked to a one-day symposium to discuss an unnerving crisis in our understanding of the universe.[1] During coffee breaks, I ran into a diverse bunch of people from all over the world: astrophysicists and cosmologists, observers and theorists, young postdocs and Nobel laureates. Some of them had spent more time on the plane than they would in the lecture room. Their mutual worry: the universe appears to be expanding too fast, and no one knows why. At the end of the meeting, Brian Schmidt, corecipient of the 2011 Nobel Prize in Physics, told me, "I'm even more puzzled after today."

Here's what astronomers and physicists alike were scratching their heads about: working from detailed studies of the cosmic microwave background, the popular ΛCDM model yields a very precise value for the current expansion rate of the universe, with an error margin of just 1 percent. However, "local" measurements, based on observations of galaxies in the relatively nearby universe, arrive at a number that is almost as precise but more than 9 percent higher. And according to Schmidt's colleague Matthew Colless, one of the organizers of the Berlin symposium, neither side has obvious weak points, even though they can't both be right.

While some scientists still think there may be an undiscovered error in either one of the two approaches (or maybe in both!), most believe the results are solid. But that doesn't mean they know how to explain the discrepancy. Even very creative minds like Harvard theorist Avi Loeb are stumped. "I tried to come up with a solution to present at the symposium," he told the audience, "but I have nothing new to report. It's not a simple problem to solve." According to Schmidt, there might be something fundamentally wrong with our interpretation of the cosmic microwave background. Or, who knows, with our current ideas about dark matter.

The determination of the expansion rate of the universe has a history of crisis and controversy. For starters, the earliest guesstimates, back in the 1930s, seemed to indicate that the universe was much younger than the Earth. And just thirty years ago, you'd be offered values that differed by a factor of two, depending on whom you asked. But cosmology has become a high-precision science, and never before has the gap between two different estimates of the Hubble constant—a measure of the current expansion rate—been so statistically significant.

In chapter 15 I explained how the expansion of empty space pushes galaxies away from each other. As a result, all cosmic distances grow by 0.01 percent in 1.4 million years, corresponding to a Hubble constant (or Hubble parameter, usually denoted H_0) of some 70 kilometers per second per megaparsec. But for many decades, the true value of the Hubble constant remained elusive. To determine it, you need to know both the cosmological recession velocity of a galaxy and its distance. In principle the recession velocity, the rate at which a galaxy's distance is increasing due to the expansion of the universe, can be found by measuring the redshift. But for a nearby galaxy—one for which it's relatively easy to measure the distance—the redshift measurement is compromised by the galaxy's real motion through space. These spatial velocities can be as high as a few hundred kilometers per second. As for remote galaxies—the ones for which any spatial motion is negligibly small compared to the cosmological recession velocity—it's frustratingly hard to measure their distances.

Over the decades, astronomers have found a solution by setting up an elaborate distance ladder to establish how far away other galaxies are. A key ingredient of this technique is a kind of star known as a Cepheid. A Cepheid is a type of variable star: its temperature rises and falls, its diameter expands and contracts, and it brightens and fades. These pulsations occur periodically; the more luminous a Cepheid is, the slower it pulsates. Henrietta Swan Leavitt at Harvard College Observatory discovered this period-luminosity relationship in the early 1900s, and it's now known as the Leavitt Law. So if you find a Cepheid in another galaxy, its observed period tells you how luminous it is, and the star's apparent brightness then reveals the galaxy's distance.

In the 1990s, using the eagle-eyed vision of the Hubble Space Telescope, a team led by Wendy Freedman (now at the University of Chicago) succeeded in identifying Cepheid variables in spiral galaxies at distances of hundreds of millions of light-years. The final results of their Hubble Key Project, published in 2001, yielded a Hubble constant of 72 km/s/Mpc, but the uncertainty in that value was some 10 percent.[2] Still, this was a huge achievement: before the launch of Hubble in April 1990, the best estimates for H_0 ranged from 50 to 100 km/s/Mpc. Moreover, the Hubble results enabled astronomers to calibrate other distance indicators that could be used farther out, where individual Cepheids can't be seen anymore.

One of those distance indicators is a type Ia supernova, which is known as a standard candle—a light source with a well-defined luminosity. By studying these stellar explosions, astronomers found that the cosmic expansion rate isn't slowing down over time, as had always been assumed, but is actually speeding up despite the mutual gravitational attraction of all matter in the universe. As we saw in chapter 16, this discovery is now seen as evidence for the existence of dark energy, the true nature of which is just as mysterious as that of dark matter.

No one realized it at the time, but the discovery of the accelerating expansion of the universe, which earned Schmidt, Saul Perlmutter, and Adam Riess the 2011 Nobel, germinated the crisis in cosmology that was the topic of the Berlin symposium. Not because the concept of dark energy is somehow deficient but because it works too well: the current expansion rate of the universe turns out to be substantially higher than theoretical cosmologists would expect on the basis of their cherished ΛCDM model.

The cosmological concordance model successfully accounts for the observed properties of the cosmic microwave background. The statistical distribution of the "hot" and "cold" spots in the CMB can only be understood if 68.5 percent of the matter-energy density of our universe is accounted for by dark energy, 26.6 percent by dark matter, and no more than 4.9 percent by ordinary, baryonic matter.[3] These cosmological parameters have now been derived so precisely that it's easy to deduce what the current value of the Hubble constant should be: 67.4 km / s / Mpc, with an error margin of less than 1 percent. (Of course, this deduction accounts for the fact that the cosmic expansion rate first slowed down because of the universe's self-gravity but is now speeding up again because dark energy started to prevail some five billion years ago.)

However, these results don't jibe with the latest local measurements of H_0 from Cepheids and supernovae. The 2001 result from Freedman's Hubble Key Project had a large enough range of uncertainty that, at first, there didn't seem to be cause for concern. But over the past decade, a team led by Riess has arrived at a much more precise calibration of the cosmological distance ladder, and at a correspondingly much more precise value for the Hubble constant. The result, as presented by Riess at the Berlin symposium: 73.5 km / s / Mpc, with an uncertainty of just 2.2 percent. "The value hasn't changed very much," he says, "but the uncertainty has come down significantly."

To achieve this high level of precision, Riess and his collaborators used a novel technique with the Hubble Space Telescope to precisely measure the distances of five Cepheid variable stars in our own Milky Way galaxy—a necessary step in accurately calibrating the Leavitt Law. Subsequently, they studied Cepheids in galaxies in which type Ia supernovae had also been observed. Using the Cepheid distances of these galaxies, the team then calibrated the standard candle properties of type Ia's. Finally, they derived the Hubble constant from observations of hundreds of supernovae in more distant galaxies, for which the redshift is a reliable measure of the cosmological recession velocity.

The two values for the Hubble constant—one from the cosmic microwave background, the other from Cepheids and supernovae— seemed as incompatible as East and West Berlin in the days of the

Cold War. "Clearly, we have not solved things today," Schmidt stoically observed in the closing session of the Berlin symposium. No walls had been torn down, and no one could think of a cosmological Checkpoint Charlie where you might try to move from one side of the divide to the other.

Eight months later, things looked even worse. In mid-July 2019, at the Kavli Institute for Theoretical Physics of the University of California, Santa Barbara, dozens of astrophysicists and cosmologists gathered for a three-day conference, Tensions between the Early and the Late Universe, coordinated by Riess and two colleagues. Riess's SHoES collaboration (an unwieldly acronym for Supernova, H_0, for Equation of State of dark energy) had published a new paper, based on more data and an even more thorough analysis.[4] Their Hubble constant result was 74.0 km / s / Mpc, with an uncertainty of just 1 percent—almost 10 percent higher than the CMB-derived value of 67.4 km / s / Mpc and with the same level of precision.

But what's more: a completely independent technique, based on gravitational lensing, arrived at a similarly high cosmic expansion rate. In Berlin, Sherry Suyu of the German Max Planck Institute for Astrophysics had already hinted at this result, but it was now backed up by a paper that would eventually be published in *Monthly Notices of the Royal Astronomical Society* in October 2020.[5] "As [the] tension between early- and late-universe probes continues to grow," the authors wrote, "we must examine potential alternatives to the standard flat ΛCDM model. This would be a major paradigm shift in modern cosmology, requiring new physics to consistently explain all of the observational data."

Suyu's approach takes advantage of the lensing process we explored in chapter 13. As we saw, the light from a remote quasar can be split into multiple images by the gravity of a massive foreground object, such as a huge elliptical galaxy. Importantly, brightness changes in the lensed quasar arrive at Earth at different moments, because each light path has its own associated travel time. So if one quasar image exhibits a certain pattern of flickers, the same pattern will be observed in another image of the same quasar with a delay of (usually) a couple of months. From this time delay—and a precise model of the mass distribution of the foreground lens—it is possible to calculate the distances traveled. Combining this with redshift measurements yields a value for the Hubble constant to a precision of a few percent.

The international HoLiCOW project led by Suyu (H_0 Lenses in COSMOGRAIL's Wellspring) kept track of brightness variations in six gravitationally lensed quasars to arrive at a value for the Hubble constant of 73.3 km/s/Mpc, with a precision of 2.4 percent—in almost perfect agreement with the SHoES value. If you take the two results together, the discrepancy with the low "cosmological" value of 67.4 km/s/Mpc has a statistical significance of more than 5 sigma (99.99994 percent confidence), meaning that there's a chance of less than 1 in 3.5 million that the mismatch is the result of some statistical fluke.

Though less precise, a number of other techniques to determine the distances of galaxies have also arrived at high values for the Hubble constant—values close to Riess's and Suyu's. The only astrophysical result presented in Santa Barbara that disagreed came from Freedman, who used a promising new method whereby the brightest red giant stars in a galaxy are taken as standard candles. Pioneered by Freedman, this technique yields an H_0 value of 69.6 km/s/Mpc—substantially lower than the other outcomes but still well outside the error margins of the ΛCDM predictions.[6]

How, then, to make sense of the divergence over the Hubble constant? Is it around 74.0 km/s/Mpc, as Riess, Suyu, and others have found, or is it 67.4 km/s/Mpc, as indicated by the cosmic background radiation? In a report on the Santa Barbara conference in *Quanta Magazine,* journalist Natalie Wolchover quotes Riess as saying, "I know we've been calling this the 'Hubble constant tension,' but are we allowed yet to call this a problem?" To which particle physicist and 2004 Nobel Laureate David Gross replied, "We wouldn't call it a tension or problem, but rather a crisis."[7]

To make matters worse, the Hubble tension (or problem, or crisis, or whatever you want to call it) is not alone. Not only is the universe expanding too fast, as compared to the predictions of the cosmological concordance model, it's also too smooth, as evidenced by recent ground-based observations.

Ever since the big bang, and despite the universal expansion of empty space, gravity has been pulling matter together in what eventually would evolve into a web of giant clusters and superclusters, interspersed with huge, empty voids. The earliest three-dimensional maps of the universe, described in chapter 6, clearly showed this uneven, lumpy distribution of galaxies. For instance, our own Milky Way galaxy is part of the so-called Local Group, which sits on the outskirts of a giant supercluster of galaxies that has the Virgo Cluster at its core. In a perfectly homogeneous universe, there wouldn't be any galaxy concentrations.

But mapping the distribution of visible galaxies only tells you about the clustering of baryonic matter. According to the ΛCDM model, baryonic matter predominantly accumulates in areas where the density of dark matter is highest, just like white foam only marks the crests of the tallest ocean waves. If you really want to know how smooth or lumpy our universe is, in order to compare it to the theoretical predictions, a 3D galaxy map won't do. You need to map the spatial distribution of dark matter instead.

One way to achieve that is by measuring cosmic shear. Briefly mentioned in chapter 13, cosmic shear is the tiny distortion in the shapes of large numbers of galaxies, caused by the weak gravitational-lensing effects of the uneven distribution of matter—both visible and dark—in the universe. Measuring shear is complicated, requiring photographic surveys that are both very wide and very deep, making them sensitive to faint sources across large portions of the celestial sky. Observations must also be conducted at multiple wavelengths, to enable redshift measurements and corresponding distance estimates of faint galaxies. Finally, there's a large number of potential systematic errors that have to be taken care of.

Despite the many hurdles and pitfalls, three international collaborations have taken up the challenge. Using the world's largest digital cameras, they have spent many years meticulously observing millions or even tens of millions of remote galaxies in huge swaths of sky.

The first of these efforts is the Kilo-Degree Survey (KiDS), featuring the 268-megapixel OmegaCAM on the European Southern Observatory's 2.6-meter VLT Survey Telescope at Cerro Paranal in Chile and led by University of Leiden astronomer Koen Kuijken, whom we met in chapter 3.[8] KiDS started in 2011 and completed its observing program in mid-2019, having covered almost 4 percent of the sky in exquisite detail.

In 2013 the Dark Energy Survey (DES) took off, with the aim of mapping nearly one-eighth of the celestial globe.[9] DES employs the 570-megapixel Dark Energy Camera on the 4-meter Victor M. Blanco Telescope at the Cerro Tololo Inter-American Observatory, also in Chile. This weak lensing survey is led by Michael Troxel of Duke University and Niall MacCrann of Ohio State University.

The deepest survey by far is being carried out with the 870-megapixel ultra-widefield Hyper Suprime-Cam on the Japanese 8.2-meter Subaru telescope on Mauna Kea, Hawai'i. Since 2014 the survey, directed by Satoshi Miyazaki of the National Astronomical Observatory of Japan, has been studying the shapes of galaxies out to distances of almost 12 billion light-years.[10]

Although the final analyses of the three programs await publication, the results so far seem to indicate that cosmic matter is distributed more homogeneously than expected. Cosmologists use the parameter S_8 as a measure of the "lumpiness" of the universe, and the value of S_8 as measured by KiDS and DES (somewhere between 0.76 and 0.78) is some 8 percent lower than the value predicted from *Planck's* observations of the cosmic microwave background (0.83). This significant discrepancy is known as the S_8 tension.

So the popular cosmological concordance model is facing more than one difficulty. To begin with, there are issues with the properties of dwarf galaxies, highlighted in the previous chapter. At a more quantitative level, measurements of the expansion rate of the universe do not agree with the model's predictions. The same is true for the large-scale homogeneity of the distribution of dark (and visible) matter. And, of course, no one has a clue as to the true nature of the model's main ingredients—dark matter and dark energy.

"The current anomalies could represent a crisis for the standard cosmological model," says Eleonora Di Valentino of Durham University. "Their experimental confirmation can bring a revolution in our current ideas of the structure and evolution of the universe," she adds.[11] However, so far no one has been able to come up with a convincing solution or a promising theoretical way to move forward. One of the problems, according to Di Valentino: if you find a way to alleviate the Hubble tension, the S_8 tension usually increases, and vice versa.[12]

Despite all these setbacks, George Efstathiou, a vocal supporter of ΛCDM, isn't convinced that there's a crisis or that cosmology is in need of a revolution. For instance, he expects—or at least, he hopes—that the Hubble tension will be alleviated by future observations made with higher precision. According to Efstathiou, there may be an issue with the calibration of the Cepheid observations by Riess's SH0ES collaboration. "I am hopeful that with a few well-chosen observations we can finally pin down the value of H_0 to better than 2 percent," he says, as opposed to the current discrepancy of almost 10 percent between local and cosmological values.[13]

At the Berlin conference in November 2018, co-organizer Matthew Colless was equally optimistic. "The nice thing about this field is that many outstanding questions will be answered in due time," he said. "Five years from now, we'll have a much clearer view." In

Artist's impression of the James Webb Space Telescope, which will be astronomers' workhorse for years to come.

particular, astronomers look forward to more precise data on stellar distances from the European astrometric satellite Gaia, to detailed supernova observations by the James Webb Space Telescope, to high-precision measurements of the cosmic microwave background by the future Simons Observatory in northern Chile, and to new comprehensive surveys of the large-scale structure of the universe at various look-back times in cosmic history.

But what if the current tensions turn out to grow stronger with time, instead of weaker? What if there's a real, persistent mismatch between cherished theory and merciless facts? Well, in science, nature always has the last word, so in that case cosmologists may need to rethink their ideas about dark energy and dark matter and be open to what's known as new physics.

In fact, many creative minds are already walking down that path.

23

Elusive Ghosts

The inhabitants of Leopoldshafen, a small town on the bank of the Rhine River in Germany, are used to heavy traffic. Trucks carrying freight from the small harbor to the nearby city of Karlsruhe are often seen driving carefully through the narrow, historic Leopold-strasse. But the convoy of Saturday, November 25, 2006, was different. Tens of thousands of curious spectators lined the roads, as a zeppelin-like structure measuring 23 meters in length and 10 meters in diameter crawled past the half-timbered houses.

It would take another twelve years before the 1,400-cubic-meter vacuum tank, one of the largest in the world, could be put to use. In fact the tank is a key component of a giant spectrometer, the heart of the Karlsruhe Tritium Neutrino experiment (KATRIN) at the the Karlsruhe Institute of Technology.[1] Using the spectrometer, physicists are now routinely studying the properties of the most elusive elementary particles they know.

It may sound like wielding a sledgehammer to crack nuts, but come to think of it, it makes a lot of sense. Neutrinos carry no electrical charge, they weigh next to nothing, and they hardly interact with other particles, since they don't feel the electromagnetic force or the strong nuclear force. As for the weak nuclear force, that one is called weak for a reason. So you need humongous experiments

The giant spectrometer of the Karlsruhe Tritium Neutrino experiment (KATRIN) moves through the narrow streets of Leopoldshafen, Germany.

to be able to study neutrinos at all, like the 50,000-cubic-meter Super-Kamiokande instrument beneath Mount Ikeno in Japan, or the one-cubic-kilometer IceCube detector in Antarctica. KATRIN is no exception, and neutrino physicists are already dreaming of even larger facilities.

Unlike quarks and electrons, the fundamental constituents of atoms, neutrinos are not part of the material world around us. They are, however, firmly embedded in the successful Standard Model of particle physics. The prediction of the neutrino's existence, in 1930 by Austrian physicist Wolfgang Pauli, even preceded the discovery of the neutron—one of the building blocks of atomic nuclei. Still, due to the particle's ghostlike nature, it wasn't until 1956 that the "little neutral one" (the Italian name had been coined in 1932 by Enrico Fermi) was actually detected.

We now know that neutrinos come in three "flavors," associated with the three types of electrons that nature has presented us with—the lightweight "regular" electron, the heftier muon, and the even more massive tau particle or tauon. Astrophysicists have also figured out that copious amounts of low-energy neutrinos were created during the big bang, while more energetic ones are produced by nuclear fusion in stellar interiors, as well as in supernova explosions. As every gee-whiz astronomy book dutifully notes, many billions of neutrinos pass through every square centimeter of your body each and every second—completely unhindered, and without leaving a trace.

But there may be more to it than the Standard Model suggests. A fourth type of neutrino could very well exist, one that's even more difficult to detect. Like the three known flavors, it would be electrically neutral. But unlike the electron neutrino, the muon neutrino, and the tau neutrino, the new species would not even be sensitive to the weak nuclear force. As a result, it would never ever interact with other particles. And unlike the other three, this "sterile" neutrino could be a heavyweight as particles go—massive enough to be a viable dark matter candidate.[2]

The first hints that there's something wrong with our classical picture of neutrinos date back to the 1960s. In the Homestake gold mine in South Dakota, the same underground laboratory that is now home to the LUX-ZEPLIN dark matter detector, physicist Raymond Davis of Brookhaven National Laboratory succeeded in catching neutrinos from the Sun—electron neutrinos produced by fusion reactions in the Sun's core. The results were confusing, though: the experiment, a 380,000-liter tank filled with dry-cleaning fluid, detected less than half the number of electron neutrinos predicted by theory.

Scientists were quick to suggest a possible solution to the solar-neutrino problem, as it became known. Suppose neutrinos could change flavor during their 8.3-minute journey from the core of the Sun to the Homestake laboratory, as had already been suggested by Italian physicist Bruno Pontecorvo in 1957. In that case, many of the electron neutrinos produced by hydrogen fusion would arrive at Earth as muon or tau neutrinos, which Davis's setup wasn't able to detect. Such neutrino oscillations were directly measured only in 1998, by the Super-Kamiokande experiment, and in 2001 by the Sudbury Neutrino Observatory in Canada.[3]

Neutrino oscillations provide a solution to the solar–neutrino problem, but they do so at a cost. According to special relativity, the particles can suffer from this weird personality disorder only when they have a certain mass, however tiny. So the Standard Model is apparently somehow flawed, since it prescribes that neutrinos are truly massless, just like photons. Also, the realization that neutrinos really do weigh something immediately raises another question: How massive are they?

That's where KATRIN comes in. By now, there's a lot of evidence that neutrinos must be hundreds of thousands of times less massive than electrons, but KATRIN is the most sensitive instrument ever to carry out the necessary measurements. Not an easy task, as you can imagine, and not something you could do with a table-top instrument. Instead the experiment fills a couple of large, vanilla-colored lab buildings, including the huge hall that is home to the zeppelin-shaped spectrometer.[4]

The principle of the KATRIN experiment is simple enough. Radioactive tritium, a heavy isotope of hydrogen, decays into helium-3, emitting an electron and an anti-electron neutrino. Energy conservation laws dictate that these two particles together can carry at most 18.57 keV of kinetic energy—that would be the case in the rare event that the helium-3 atom ends up with no kinetic energy at all. So if you measure the energy distribution of the electrons emitted by the decaying tritium, you would expect that the most energetic ones have a kinetic energy just shy of this peak value. The minute difference then equals the kinetic energy of the accompanying antineutrinos, from which it's easy to calculate their mass—which equals the mass of *regular* electron neutrinos.

Simple in principle, but extremely difficult in practice, partly because tritium is lethal stuff; the Tritiumlabor Karlsruhe is the only European laboratory licensed to work with the highly radioactive gas. The tritium source has to be cooled down to just 30 degrees above absolute zero, while extremely strong superconducting magnets—about a hundred thousand times stronger than the Earth's magnetic field—are necessary to guide the electrons into the spectrometer. It's quite an avalanche, by the way: about hundred billion electrons enter the giant vacuum tank each and every second.

Once in the tank, the negatively charged electrons have to travel "upstream" against a powerful electric field of some 20,000 volts. Most of them are slowed down, stopped, and dragged back to where they came from. Only the most energetic ones—maybe just one in a few trillion—are able to reach the sensitive detectors at the opposite end of the spectrometer. By precisely tuning the electric field strength, physicists are able to measure the number of electrons at various energies close to the peak value of 18.57 keV.

Apart from many engineering problems, the KATRIN project also faced a logistical challenge. The spectrometer was built at MAN DWE GmbH, a German steel construction company located in Deggendorf, some 400 kilometers east of Karlsruhe. But since the device was much too big for road transport, it had to take an 8,600-kilometer detour over water. In fall 2006, the 200-ton, whale-sized behemoth traveled down the Danube river, across the Black Sea, through the Bosporus, and into the Mediterranean. Next it sailed through the Strait of Gibraltar, up the Atlantic coast, into the English Channel, all the way to the port of Rotterdam. From there, it went along the Rhine to Leopoldshafen, for its final—and attention-grabbing—road trip of just seven kilometers.

In September 2019, at the sixteenth International Conference on Topics in Astroparticle and Underground Physics in Toyama, Japan, KATRIN physicists presented the results of their first science run, indicating that electron neutrinos must weigh less than 1.1 eV (as compared to the electron mass of 511,000 eV).[5] Future measurements are expected to probe masses as low as 0.2 eV. Meanwhile, work is underway for an upgrade that would enable KATRIN to detect sterile neutrinos—which, again, do not interact at all with other particles—if they exist.

So why do physicists think there might be an as-yet-undiscovered type of neutrino? Basically, there are two theoretical reasons. First, the existence of a massive neutrino would naturally explain why the three known neutrino flavors are so incredibly lightweight. This has to do with the complicated concept of neutrino mixing, according to which a neutrino is in fact an ever-changing ("oscillating") combination of various "mass eigenstates"—if there is a fourth, much more massive sibling, it is much easier to understand why the electron, muon, and tau neutrino are almost massless, but not quite.

The second reason is the remarkable fact that all known neutrinos are "left-handed" particles. The handedness of an elementary particle is related to its spin direction relative to its direction of motion. Quarks and electrons (as well as muons and tau particles) come in both varieties, but no one has ever seen a right-handed neutrino, which seems a bit weird. Unless, for some reason, right-handed neutrinos are sterile, making them all but impossible to detect.

And of course, there is the exciting prospect that sterile neutrinos might solve the mystery of dark matter. As we saw in chapter 11, "normal" neutrinos are not massive enough to double as dark matter particles: because of their high velocities (astrophysicists call them "hot"), normal neutrinos would not have been able to clump into the galaxy-sized dark matter halos that populated the early universe. But if sterile neutrinos happen to weigh in at a few thousand eV, they would be the perfect alternative to WIMPs. Moreover, since sterile neutrinos are "warmer" than the canonical cold dark matter that physicists have fruitlessly been hunting for decades, some of the problems that the ΛCDM model is facing (see chapter 22) might disappear like snow in sunlight.

At present, though, the existence of sterile neutrinos is still heavily debated. Some neutrino experiments have provided indirect hints of the weird particle, but these results are incompatible with others, including data from the IceCube detector. Moreover, in October 2021 researchers with the MicroBooNE experiment at the University of Chicago's Fermilab announced they didn't see any evidence for the particle's existence. Future observations—possibly from the planned upgrade of KATRIN or from the Deep Underground Neutrino Experiment that is taking shape at the Sanford laboratory in South Dakota—may settle the issue. But, for now, sterile neutrinos exist only in the minds of inventive theorists.

The same is true of yet another potential dark matter candidate particle, the axion. In fact, the axion saga contains many parallels with the sterile-neutrino story: complicated theoretical reasons to believe in the particle's existence, an intriguing possibility that it could solve the dark matter riddle, tentative hints of a possible detection, ongoing searches, and no final answer yet. In both cases, some scientists feel that we are on the verge of a long-awaited breakthrough, while others see the increasing interest in sterile neutrinos and axions as dark matter candidates as evidence that desperate physicists and cosmologists are grasping at straws while the decades-long search for WIMPs continues to turn up empty.

The rationale for believing in the existence of axions has to do with antimatter. You would expect that the incredible energy of the big bang produced similar amounts of matter and antimatter particles. For some reason, however, everything in the current universe consists of normal matter—there are no antimatter galaxies, stars, planets, or living organisms, at least not that we know of. Since matter and antimatter particles annihilate upon running into each other, there must have been a teeny-weeny bias in the laws of physics: for every billion antimatter particles, nature apparently created a billion plus one particles of normal matter. After the great annihilation that played out shortly after the big bang, this minute one-part-in-a-billion residue was all that was left to clump together into galaxy clusters, planetary systems, and people. We owe our existence to this tiny asymmetry.

In 1964 physicists indeed discovered a tiny difference in the way the weak nuclear force acts upon matter and antimatter. In particular, they found that the metamorphosis of particles known as neutral kaons into their antiparticles (a transformation mediated by the weak force) is slightly less probable than the same process occurring in reverse. This rather surprising property of nature is technically known as CP-symmetry violation, where C and P denote charge and parity, respectively. It turns out that the effect is much too small to explain the matter / antimatter asymmetry of the universe all by itself, but if the strong nuclear force also violates this fundamental symmetry of nature, the mystery might be solved.

And here's the problem that axions could help alleviate. According to the Standard Model of particle physics, CP violation is allowed to take place in the strong interaction, just as it is in the weak interaction. But despite extensive, dedicated searches, it has never been seen to occur—a worry known as the strong-CP problem. It is as if the effect is forcefully suppressed, possibly by some new field, as proposed by Roberto Peccei and Helen Quinn in 1977.[6] And if that field exists, it should have an associated particle: the invisible axion. (This is somewhat akin to the situation with the Higgs field, which was proposed to explain how elementary particles achieve mass. The associated Higgs particle was discovered at CERN in 2012.)

If "axion" sounds like a washing powder brand, that's because American physicist Frank Wilczek named it after a presoak and detergent booster he came across in an advertisement in 1978. Just as the packaging promised "safe whitening brightening power for all your wash," the new particle would clean up the strong-CP problem in physics. (Incidentally, the Greek "axios" means "worthy" or "deserving.")

More than forty years after Wilczek coined the particle's catchy name, still no one knows if axions really exist. But if they do, they must be extremely lightweight—much less massive even than "regular" neutrinos. Expressed in energy units, where a proton mass equals about 1 GeV and an electron weighs in at 511 keV, the mass of the axion would probably have to be measured in microelectronvolt.

So what about their dark matter candidacy? Well, axions theoretically are stable, they do not have an electrical charge, and they are about as interaction-shy as sterile neutrinos—three defining properties of any dark matter particle. Moreover, if axions exist, theory predicts them to be incredibly numerous. Each and every cubic centimeter of the cosmos would contain an average of tens of trillions of axions. So despite their incredibly tiny mass, they could very well make up the bulk of our universe, thanks to their unimaginable abundance.

But wait—if the mass of the axion is so small, doesn't that mean that it, too, moves at relativistic speeds, just like the common neutrino? As we saw, neutrinos cannot constitute the universe's dark matter because they are hot and unable to easily clump into small structures. So why would axions fare any better? Well, that's because they formed very differently. Unlike neutrinos (and WIMPs), axions were not in thermal equilibrium with other particles at the time of their formation—which is when individual quarks combined into composite particles, including protons and neutrons. Instead, quantum physics predicts that axions formed a so-called Bose-Einstein condensate, where large groups of identical particles behave as if they were just a single particle. The result is that they are cold particles, despite their tiny mass.

OK, so axions might exist, and they might solve the mystery of dark matter. Great. But how can you possibly hope to detect them?

The good news is that axions are not completely immune to the electromagnetic force, even though they don't have an electrical charge themselves. A strong magnetic field can help to turn an invisible axion into a visible photon, and vice versa, where the wavelength of the photon—a particle of light—is directly related to the mass of the axion. This realization has led to a number of experiments. For instance, the CERN Axion Solar Telescope (CAST) in Geneva has been looking for relatively massive solar axions since 2003.[7] If such axions are produced in the Sun's core, by the interaction of energetic X-rays with charged particles, they would arrive at Earth in huge numbers, and a powerful magnetic field could turn some of them back into X-ray photons. CAST uses a decommissioned superconducting test magnet from CERN's Large Hadron Collider in combination with sensitive X-ray detectors to look for this process to occur—so far without any success.

At the DESY research laboratory in Hamburg, Germany, physicists of the ALPS experiment (Any Light Particle Search) use a different technique: they try to shine an infrared laser through a wall.[8]

Under normal conditions, no single photon would be able to pass through the light-blocking barrier, but by applying a strong magnetic field, some of them might turn into axions (or axion-like particles) before hitting the wall. The axions hardly interact with anything, and when they end up on the other side of the barrier, the same magnetic field can turn a few of them back into infrared photons, ready to be detected. It's a neat setup, but just like CAST, it hasn't found any evidence for axions yet.

By far the most exciting axion detector—at least for dark matter hunters—is based at the University of Washington in Seattle. The Axion Dark Matter eXperiment, or ADMX, follows on an idea first proposed by Pierre Sikivie of the University of Florida.[9] ADMX tries to detect the very-low-mass axions that could be abundant in the Milky Way's halo, which is why Sikivie coined the name axion haloscope for the instrument.[10] It's basically a cylindrical vacuum vessel (a so-called resonant cavity) cooled down to just above absolute zero and surrounded by an 8-tesla magnet—2,000 times stronger than a typical fridge magnet. The strong magnetic field would occasionally turn halo axions into microwave photons. The expected signal (at most 10^{-21} watts, also known as 1 zeptowatt) should be detectable through the use of superconducting quantum amplifiers.

Conceived in the 1990s, ADMX is now a collaboration of researchers from twelve institutes in the United States, the United Kingdom, and Germany, largely funded by the US Department of Energy. The axion haloscope can be tuned to a range of microwave frequencies, corresponding to axion masses between 1 and 40 microelectronvolt (μeV). As of this writing, only values between 2.66 and 3.31 μeV have been conclusively ruled out, but eventually, the full plausible dark matter mass range will be explored and scrutinized.

While dedicated searches for axions have been unsuccessful so far, Elena Aprile's XENON collaboration provided an intriguing twist to the story. During XENON1T's first science run, between February 2017 and February 2018, the instrument detected a small excess of low-energy events: a few dozen minute signals due to interactions not with xenon nuclei, as you would expect from WIMPs, but with xenon electrons. Such electron recoil events are generally produced as a result of background noise, like the radioactive decay of radon and krypton atoms, but after a careful analysis, the team wasn't able to account for the observed events at energies between 2 and 3 keV.

One possible explanation, explored in an October 2020 *Physical Review D* paper from the XENON team, is the existence of axions—not of the cold dark matter type, to be sure, but fast-moving axions produced in the core of the Sun, more or less similar to the axions that the CAST experiment is looking for.[11] If confirmed, the discovery of fast-moving axions would be momentous, but the truth is that a much more mundane explanation cannot be ruled out: an extremely small, unaccounted-for overabundance of radioactive tritium atoms in the liquid xenon—just a few atoms per kilogram— would produce a similar excess. By the time this book is published, the issue probably will have been settled by the much more sensitive data from the larger XENONnT experiment and its American competitor LUX-ZEPLIN.

Both sterile neutrinos and axions were first proposed decades ago in attempts to solve nagging problems in particle physics. Soon enough, desperate dark matter hunters considered them potential candidates for the invisible gravitating mass in the universe. But the existence of both particles is still highly speculative, and within another couple of decades, they may well end up in the ever-expanding graveyard of theoretical dead ends and dismissed hypothetical particles.

Time to start looking for even weirder alternatives.

24

Dark Crisis

Amsterdam is a ghost city. The square in front of the nineteenth-century central railway station is all but empty. There are no back-packers cycling along the canals. No drunk British hooligans strolling through the Red Light District. No American tourists waiting in line for the Van Gogh Museum. Most schools, theaters, and shops are closed. People work from home, avoiding personal contact.

Despite the strict lockdown measures issued by the Dutch government, I am meeting theoretical physicist Erik Verlinde in person, in his small office at the eerily quiet University of Amsterdam.[1] We are wearing face masks. We don't shake hands. We keep our distance. As most scientists do, Verlinde takes the invisible coronavirus very seriously. He doesn't believe in dark matter, though.

It's December 2020, and I'm holding my first personal interview in six months. The COVID-19 pandemic has thrown a wrench into my book research. The biennial UCLA Dark Matter Conference, planned for late March, was canceled. Most of my interview appointments were rescheduled as Zoom meetings. I was unable to visit the LUX-ZEPLIN experiment in South Dakota or the Dragonfly Telephoto Array in New Mexico. IDM 2020, a conference on the identification of dark matter in Vienna, was held online, with a reduced number of talks. And no, I couldn't travel to the

China Jinping Underground Laboratory to check out the PandaX detector.

The coronavirus also left its mark on the progress of dark matter research. Astronomical observatories and physics labs, including CERN and Gran Sasso, had to be closed down. International projects faced delays due to travel restrictions and quarantine regulations. Researchers got ill, and some died. The decades-long search for the true nature of dark matter almost ground to a halt, except that theorists like Verlinde never stopped thinking about novel ways to solve the crisis.

Not the COVID-19 crisis, mind you, but the dark matter crisis. Which, incidentally, bears a number of resemblances to the corona pandemic. In the case of dark matter, the contagion is worry: the growing concern among researchers that they've been barking up the wrong tree. Dark matter may not consist of WIMPs, after all. Even the ΛCDM model may be faulty. Perceived certainties fall away, there's room for—and need of—novel ideas, and we probably have to prepare for a new normal, even though no one knows where we're heading. Sounds familiar.

While old hands like CERN's John Ellis and XENON lead Elena Aprile still think our solar system may plough through a sea of weakly interacting massive particles that might be detected by some future underground experiment, a younger generation of scientists—some of whom were not even born when the concept of WIMPs was introduced—are ready to bid the idea farewell. "The WIMP model may not be dead," says Caltech theorist Kathryn Zurek, "but it's certainly on life support. If I'd have to bet on a dark matter candidate, it wouldn't be on the WIMP."[2]

Likewise, theorist and author Sabine Hossenfelder of the Frankfurt Institute for Advanced Studies in Germany feels that WIMP searches have always been badly motivated.[3] As we saw in chapter 10, the big bang theory predicts a "relic density" for the hypothetical WIMPs that beautifully matches the mass density of cold dark matter. But, as Hossenfelder sees it, "this WIMP miracle shouldn't have been taken seriously. Beauty is not a scientific argument. It should never have been made." She strongly believes that, in general, the predilection for mathematical beauty leads physicists astray, whether in their quest to solve the dark matter mystery or their search for an all-encompassing theory of everything.

Numerical arguments and coincidences often play an important role in the development of new physical theories. For instance, some theorists wonder about the fact that the mass density of nonbaryonic dark matter in the universe is not dramatically different from the mass density of atomic nuclei. Sure, there's five times more dark matter than "normal" matter, but it's the same order of magnitude, while there's no obvious reason why the difference wouldn't be a factor of a million. So maybe, these theorists argue, nature is telling us something—maybe there's something to explain here. That line of thought has led to the idea of asymmetric dark matter, in which dark matter particles are not their own antiparticles, as could be the case for WIMPs, but instead are subject to the same particle / antiparticle asymmetry as baryons. If so, it's not surprising that the total amounts of baryonic matter and dark matter are comparable.

Because the hunt for WIMPs has been fruitless so far, says Hossenfelder, scientists are starting to reorient a bit, broadening their view to include more speculative ideas. Indeed, if you browse through a couple dozen issues of *New Scientist* or check the abstracts of papers published on the arXiv preprint server, there's an overwhelming number of crazy concepts and wild theories. Even some ideas that were written off long ago are back on the table.

Take primordial black holes, for example. These tiny knots of strongly curved spacetime, first proposed in the 1970s by Bernard Carr and Stephen Hawking, were soon suggested as dark matter candidates but fell out of favor when they didn't show up in microlensing surveys.[4] However, like old rock stars, they are now experiencing something of a comeback.

To be clear, we are talking about a very specific kind of black hole, not just any. Upon hearing about the riddle of dark matter, many people immediately think of black holes as the most likely solution. After all, black holes are invisible, massive, stable, and mysterious—what more do you need? The truth, however, is that regular black holes—both the relatively lightweight ones that are left when massive stars go supernova, and the supermassive ones in the cores of galaxies—cannot possibly constitute the dark matter in the universe. First of all, from what astrophysicists know about the evolution of galaxies and stars, it follows that at most one-hundredth of a percent of all the mass in the universe is now contained in black holes. But more importantly, all these black holes formed from baryonic matter, in the course of 13.8 billion years of cosmic history. They belong to the baryonic, 4.9 percent slice of the cosmic pie described in chapter 16.

Primordial black holes, on the other hand, may have formed from intense fluctuations in the very fabric of spacetime during the universe's birth—maybe even during the brief period of exponential expansion known as the inflationary era. That was well before the origin of atomic nuclei and before the relative amounts of baryonic and nonbaryonic matter were established. An ocean of primordial black holes, each about the size of an atomic nucleus and with the same gravitational pull as a small planet, could very well explain the dark matter mystery. At least, that's what some physicists thought.

But when the MACHO and EROS experiments failed to see evidence of massive, compact objects in the Milky Way's halo (see chapter 14), primordial black holes lost their popularity as dark matter candidates.

Quite recently, however, primordial black holes have been revived. Theorists have described the possibility that the universe has created much more massive versions of the enigmatic objects, packing in the mass not of a small planet but perhaps dozens of Suns.[5] If the dark matter in the universe is composed of 30-solar-mass primordial black holes, they would be more widely separated from each other than would the smaller hypothesized primordial black holes of yore, and microlensing surveys may just have missed them.

It's a perfect example of what theoretical research on dark matter has become. You resurrect an old idea—or, better still, come up with a completely novel one—tweak it until it no longer contradicts accepted scientific wisdom and observational evidence, make sure it is internally consistent, and you're in business. No need to come up with an experimental justification anymore; as long as your proposal isn't completely metaphysical and has the potential of resolving the dark matter crisis, it has a decent chance of ending up in *Physical Review Letters* or *The Astrophysical Journal*. And the longer it survives, the more convinced you may become that you're on the right track.

Many theorists would say that this is indeed the best way to go forward: to leave no stone unturned. In scientific jargon, to fully explore the available theoretical parameter space. Be that as it may, this approach also yields a plethora of far-fetched, speculative notions, the vast majority of which are necessarily erroneous—after all, there's only one truth.

But this is how physics works, and over the years scientists have come up with a wide range of exotic concepts. One of them is fuzzy dark matter, the topic that Jerry Ostriker is currently working on.[6]

Fuzzy dark matter would consist of particles with an incredibly small mass of 10^{-22} eV. Because of this infinitesimal mass, the particle's associated quantum wavelength—its "fuzziness" due to quantum uncertainty—would be thousands of light-years. This means that, on the scale of galaxies, the hypothetical stuff behaves very differently from any other form of particle dark matter, solving many of the problems discussed in chapter 21. Needless to say that there's no chance whatsoever of directly detecting particles that weigh even less than a trillionth of a trillionth of the mass of an electron—which, incidentally, may be good for the theory's longevity.

Other novel ideas include decaying dark matter, which has recently been proposed in an effort to alleviate the so-called Hubble tension—the fact that the universe appears to be expanding significantly faster than you would expect by extrapolating from the cosmic microwave background data.[7] If dark matter particles gradually decay into some form of "dark radiation," their total gravitational pull would lessen over time. This reduction in gravitational pull, combined with the accelerating effect of dark energy, would allow cosmic expansion to speed up enough to explain the relatively high rate of expansion that astronomers have observed.

Of course, if dark matter decays into dark photons, it must be subject to some unknown force. Indeed, some physicists speculate that there's not just one type of dark matter particle but a whole multicomponent "hidden sector" of dark particles, dark forces, and dark force–carrying bosons, also known as dark photons. After all, the known subatomic world—the Standard Model—is utterly complicated and inelegant, so why would you expect the dark, hidden side of nature to be simple and minimalistic? As a bonus, a populated hidden sector, with its own diversity of particles and interactions, provides a wealth of new possibilities to explain the weird observational properties of the universe.

A wider variety of hypothetical particles and forces is also good news for inventive experimentalists, who design experiments that may provide empirical evidence to support theorists' wild ideas. One of these new experiments is a detector called FASER (ForwArd SEaRch experiment), which will start operations at CERN in 2022.[8] "During the next run of the Large Hadron Collider, which will last for three years, we expect FASER to detect about one hundred dark photons," says CERN staff scientist Jamie Boyd.[9]

In summer 2019, when I visited CERN, FASER was still very much in its early construction phase—it had only been approved a few months before. Some of the detector's massive scintillators (spare parts from an older particle physics experiment at CERN) were still

FASER (ForwArd SEaRch experiment), a new detector installed at CERN to search for dark photons or other relatively long-lived particles.

stored in Boyd's office. At the far end of a long, dimly lit corridor, way past the small office where Tim Berners-Lee developed the World Wide Web in the early 1990s, Boyd showed me around a machine shop where technicians were running tests on sensitive spectrometer modules for FASER—again, hardware that was designed and built for other experiments.

Short-lived particles produced by the smash-up of protons in the Large Hadron Collider are caught by the collider's large detectors, notably ATLAS and CMS. But what if dark photons would also occasionally be produced in the aftermath of the collisions? They would fly off along a straight path tangent to the collider's circumference and would only decay into high-energy electron / positron pairs after a few hundred meters, well outside the main particle detectors.

Based on an idea developed by physicist Jonathan Lee Feng of the University of California Irvine, the $2.5 million FASER experiment that Boyd is working on is constructed in an old, deserted tunnel segment that just happens to lie in the right place, some 480 meters away from the ATLAS detector. Within a few years, FASER may find supporting evidence for the existence of dark photons or other relatively long-lived particles. On the project's website, it's called the BSM program, for Beyond the Standard Model.

Yet another far-fetched idea is superfluid dark matter, first conceived by Lasha Berezhiani, now at the Max Planck Institute for Physics, and Justin Khoury of the University of Pennsylvania.[10] Just like water can exist in different phases (as vapor, liquid, and ice), extremely lightweight, axion-like dark matter particles might exist in multiple phases, too. At very low pressures, the mysterious stuff would behave like a normal gas of particles, interacting only through gravity. But in denser regions, like the dark matter halos that eventually give rise to galaxies, some form of self-interaction would turn

the particles into a superfluid, with very different properties, somewhat akin to the frictionless behavior of superfluid liquid helium.

Speculative? Yes, absolutely. But the attractive thing about the
theory is that the interaction of superfluid dark matter with normal
baryonic matter would produce a new force that acts more or less
like gravity. It's why Hossenfelder is so fond of the idea. "People
have been trying to solve the dark matter mystery either with one
type of particle, or with some form of modified gravity," she says.
"By far the most neglected option is a combination of both."

As noted before, ΛCDM—with particle dark matter—works very
well at cosmological scales, but it fails on the scale of individual galaxies, where modified Newtonian dynamics (see chapter 12) is
much more successful. However, says Hossenfelder, MOND cannot
be the answer, as it doesn't explain the properties of galaxy clusters
or the cosmic microwave background. Moreover, there's no viable
relativistic version of the theory. But a new force, exerted by superfluid dark matter, would look like some form of modified gravity
on galactic scales. That's why Hossenfelder calls it an "impostor
field."

"The observational data are telling us that we do have a regime
where things respond differently," she says. "It's a mistake to regard
particle dark matter and modified gravity as two competing theories, each of which has to be made to fit all of the data." After all,
as she points out in her popular "Backreaction" blog and YouTube
channel, you wouldn't try to tweak the Bernouilli equations, which
describe the behavior of flowing fluids, to explain the properties
of ice.[11] If a phase transition occurs, you're really talking about two
very different dark matter regimes, each with its own characteristic description.

Sterile neutrinos, axions, asymmetric dark matter, fuzzy dark
matter, multicomponent dark matter, self-interacting dark matter,

superfluid dark matter—new theoretical flights of fancy appear to pop up faster than mutant versions of a deadly virus. Is this really a sign of progress? An indication that we're finally closing in on the one and only solution to the largest mystery in astrophysics? Or is this the hallmark of a real crisis, with scientists desperately in search of an answer that keeps eluding them? Maybe we are like the blind men in the Hindu fable, trying to explain everything in terms of a wall, a spear, a snake, a tree, a fan or a rope, without seeing the elephant for what it really is?

At the University of Amsterdam, Erik Verlinde goes a step further. He doesn't believe there's an elephant at all. According to his theory of emergent gravity, dark matter doesn't exist.[12] Instead, what we perceive as the gravitational effect of mysterious dark stuff is really the interaction between normal matter and omnipresent dark energy, while dark energy results from the thermodynamic properties of the cosmological horizon—the "edge" of the observable universe. If this sounds unintelligible, don't worry. Few people are familiar with emergent gravity, and even Verlinde himself admits there are lots of loose ends.

Together with his twin brother Herman, who is now a string theorist at Princeton University, Erik studied physics at the University of Utrecht, where Dutch Nobel Laureate Gerard 't Hooft kindled his interest in black holes. Back then, the existence of these gluttonous cosmic enigmas was still debated, but physicists like Jacob Bekenstein, Hawking, and 't Hooft had carried out lots of theoretical research on black holes. In particular, Hawking demonstrated that they must emit a tiny amount of radiation, due to quantum effects near the so-called event horizon, the distance at which it becomes impossible to escape the gravity of a black hole. Bekenstein, meanwhile, showed that black holes must have a certain amount of entropy (or disorder), proportional to the surface area traced by the event horizon. And the relationship between the thermodynamics and gravity of black holes was a key ingredient of 't Hooft's holographic principle, which is too complicated to describe here in detail.

The surprising link between thermodynamics and gravity, as evidenced by the mathematical properties of a black hole's event horizon, led Verlinde to his theory of emergent (or entropic) gravity. Building on the work of Ted Jacobsen (now at the University of Maryland), Verlinde proposes that the equations of Einstein's general theory of relativity—our best description of gravity—can be deduced from some underlying microscopic properties of spacetime, just as the laws of thermodynamics that were formulated by Ludwig Boltzmann in the 1880s can be deduced from statistical mechanics: the collective microscopic behavior of large numbers of particles.

In other words, gravity is not comparable to the other known forces of nature—tellingly, it is not part of the Standard Model of particle physics—but naturally emerges at a macroscopic level from the more fundamental characteristics of spacetime, just like the temperature of a gas is a macroscopic physical quantity that emerges from the underlying properties and behavior of trillions of molecules. "If you understand statistical mechanics, you understand thermodynamics," says Verlinde. "Likewise, if we would fully understand spacetime, we would understand gravity."

And there's more. According to Bekenstein and Hawking, the event horizon of a black hole—the "surface" from beyond which no information can reach us—has a certain temperature, corresponding to its entropy. According to Verlinde, the same must then be true of the cosmological horizon. And just as the temperature of a black hole's event horizon has an effect on the visible universe, in the form of Hawking radiation, the thermodynamical properties of the cosmological horizon leave their mark in the visible universe in the form of dark energy.

So what about dark matter? Unnecessary, says Verlinde. What astrophysicists interpret as the gravitational effect of dark matter is actually the result of an interaction between this completely new description of dark energy and the baryonic matter in the universe. Interestingly, Verlinde's preliminary calculations are pretty much in line with the results from the rather ad hoc theory of modified Newtonian dynamics. "I certainly can't explain everything," he says, "but that doesn't necessarily mean that the idea is wrong."

So far, his theory is very much work in progress, and it doesn't have many adherents yet. Part of the problem might be that it challenges so many cosmological orthodoxies. For instance, as Verlinde dryly notes, "the whole concept of a big bang doesn't really fit well with my idea." But who knows, in times of crisis, a great reset may be the best way out, just like the COVID-19 pandemic is demanding a global reset of economies, societies, and health-care systems.

When I leave the University of Amsterdam's Institute for Theoretical Physics, the Sun is already setting. It's getting dark outside, and it's raining. I feel depressed—no one knows how long the current lockdown will last. Again, there's a bleak resemblance with the dark matter quandary in cosmology. Will it ever end?

But five days later, things look less gloomy. A UK grandmother is the first person in the world to be given the newly developed Pfizer / BioNTech vaccine. As more and more people are vaccinated against the coronavirus over the next few weeks, I realize that there's a way out of every crisis, and that science never gives up.

As Robert Kennedy once said, "The future is not a gift; it is an achievement."

25

Seeing the Invisible

On the eastern facade of a huge, white building is a small door. On the door is a tiny sign saying, "Salle Blanche Euclid." Beneath it someone has taped a piece of paper with a handwritten request: "Bien fermer la porte—Merci" ("close the door tightly—thanks"). Inside the building, the European Space Agency's next space telescope is taking shape.

From the historic Place du Capitole in the bustling center of Toulouse, it's forty minutes by public transport to Airbus Defence and Space, on the Rue des Cosmonautes.[1] Ten minutes later, project manager Laurent Brouard is showing me around in the high-tech clean room where two cameras—the Visual Imager and the Near-Infrared Spectrometer and Photometer—are being tested. The cameras are part of the agency's Euclid mission, a six-year operation during which these sensitive instruments will image billions of galaxies across one-third of the celestial sky, out to distances of 10 billion light-years. The goal: to map the geometry of the universe, in an attempt to better understand dark energy and dark matter.

The $800 million Euclid mission, named after the Greek founder of geometry, is slated for launch in late 2022.[2] "We have experienced a few months of delays due to the COVID-19 pandemic," Brouard tells me in his heavy French accent. Fully dressed up like

Artist's impression of the European Space Agency's Euclid spacecraft. Scheduled for launch in late 2022, Euclid will study the shapes and spatial distribution of billions of galaxies.

workers in an intensive-care unit, we enter a dark, ultraclean part of the "salle blanche," where Euclid's silicon-carbide payload module is being assembled. Technicians use lasers and interferometers to check the micrometer-level alignment of the space telescope's silver mirrors, the largest of which measures 1.2 meters across.

"We're about to put the telescope's baffle in place," Brouard says. "You're one of the last people to actually see the primary mirror." It's a strange thought: a few years from now, this shiny surface, polished to a precision of some 50 nanometers, will reflect billion-year-old photons from remote galaxies, enabling astronomers to map the 3D distribution of dark matter and the expansion history of the universe. Weak gravitational lensing, warped galaxy images, baryon acoustic oscillations, the accelerating expansion of empty space—I wonder what Euclid of Alexandria would've thought about it all.

Euclid—the space telescope—builds on the experience gained from ground-based projects like the aforementioned 2dF Galaxy Redshift Survey, Sloan Digital Sky Survey, Kilo-Degree Survey, Dark Energy Survey, and Hyper Suprime-Cam Survey. But from its vantage point in outer space, 1.5 million kilometers behind the Earth as seen from the Sun, the European space observatory will not be hampered by atmospheric turbulence. Moreover, it can study the distant universe 24 / 7, making it much more efficient than Earth-bound telescopes.

It won't be long before Euclid gets company from its American counterpart. The Nancy Grace Roman Space Telescope—the mission formerly known as WFIRST, for Wide-Field Infra-Red Space Telescope—will launch in the mid-2020s.[3] Named after Nancy Grace Roman, who was NASA's first chief of astronomy and "mother of Hubble," the new telescope sports a primary mirror as large as that of the famous Hubble Space Telescope (2.4 meters) but with a much wider field of view. Its 300-megapixel camera will study less cosmic real estate than Euclid but in more depth and at a wider range of wavelengths, providing observations of weak lensing, cosmic shear, and baryon acoustic oscillations.

The same kind of observations will also be the focus of the Legacy Survey of Space and Time, carried out by the 8.4-meter Simonyi Survey Telescope at the Vera C. Rubin Observatory in Chile (see chapter 6). And all these future projects will benefit heavily from the new Dark Energy Spectroscopic Instrument (DESI) mounted on the 4-meter Mayall Telescope at Kitt Peak National Observatory near Tucson, Arizona. While color observations by Euclid, Roman, and the Rubin Observatory will provide a so-called photometric redshift and a rough estimate of a galaxy's distance, DESI's detailed spectroscopic measurements will yield precise redshifts and distances for tens of millions of remote galaxies and quasars across one-third of the sky.[4]

By combining the data from various ground- and space-based facilities, astronomers can create a three-dimensional map of a substantial fraction of the observable universe. Studying the growth of baryon acoustic oscillations over time—the patterns imprinted on the cosmic mass distribution some 380,000 years after the big bang—makes it possible to deduce the expansion history of the universe, which in turn tells you about the behavior of dark energy. Meanwhile, using weak gravitational lensing and cosmic shear to measure the subtle ways in which the shapes of distant galaxies are distorted by the uneven distribution of mass between those galaxies and Earth discloses the whereabouts of dark matter through space and time.

As of this writing, Euclid is being readied for final testing, work on the Roman Space Telescope is about to commence, construction of the Vera Rubin Observatory is nearing completion at Cerro Pachón, and DESI is all set to start its five-year spectroscopic survey. In addition, the universe is monitored by sensitive gravitational wave detectors, the James Webb Space Telescope should be in space by the time this book is published, European astronomers are building the Extremely Large Telescope, and the two-continent Square Kilometre Array—the largest radio observatory ever—is under construction in Australia and South Africa. It's hard to keep track of all the new projects, as astrophysicists miss no opportunity to wrangle new information about the dark side of the universe out of every photon they can lay their hands on.

But learning about the temporal evolution of dark energy and the spatial distribution of dark matter doesn't tell you what kind of stuff we're dealing with. We're still studying footprints in the mud, in ever-increasing detail, without actually identifying the invisible man who is leaving them. If you want to discover the true nature of dark matter—if you really want to "see" the invisible—terrestrial telescopes and space-based observatories are not enough, no matter how powerful they are. The future of dark matter research will be determined (or achieved, if you will) as much by particle physicists as by astrophysicists. "The new guiding principle should be "no stone left unturned," according to particle physicists Gianfranco Bertone and Tim Tait in a 2018 *Nature* review article.[5]

In their subterranean caves and tunnels, dark matter hunters follow the same strategy as sky-watching astronomers: if you're looking for something but can't find it, just build a bigger, more sensitive, and more efficient instrument. It worked with neutrinos and quarks; it worked with black holes, extrasolar planets, and gravitational waves; it worked with the Higgs boson—there's no obvious reason it won't work with dark matter. Who knows, the breakthrough discovery might lie just beyond the horizon of our current technological limits.

It's certainly what Laura Baudis is hoping for, although she realizes that "nature doesn't really care about our hopes," as she says.[6] A particle astrophysicist at the University of Zürich, Baudis is the spokesperson of the DARWIN collaboration, a group of some 170 scientists from more than thirty institutes in Europe and the United States. Their aim: to build the ultimate liquid-xenon dark matter detector, dwarfing the currently state-of-the-art XENONnT in Italy, LUX-ZEPLIN in South Dakota, and PandaX-4T in China. If approved, says Baudis, construction of the $150 million instrument could start in 2024, most likely in the Gran Sasso laboratory, with the first scientific results expected in 2027.[7]

Baudis remembers sitting in the Zürich airport one day and noticing a plane belonging to a small regional airline called Darwin.

"I immediately liked the name," she says. "Back then, we weren't yet sure whether the new detector would use liquid xenon or liquid argon, and DARWIN became an acronym for DARk matter WIMP search with Noble liquids." (The airline went bankrupt in late 2017, by the way.) Loaded with a whopping 50 tons of liquid xenon, DARWIN will have unprecedented sensitivity, enough to register dark matter particle interactions that are too rare for current experiments to detect. "It would be sort of crazy not to cover this gap," Baudis told *Nature* reporter Elizabeth Gibney.[8] "Future generations may ask us, why didn't you do this?"

Eventually, however, neutrinos will close the door on the direct detection of WIMPs. Neutrinos from the Sun and "atmospheric neutrinos" produced by cosmic rays constitute a low but persistent background signal that no single technology can shield against. If the long-sought dark matter interactions are so rare that they are washed out by this neutrino background, physicists will "hit the neutrino floor," as they call it, and direct detection of WIMPs will be impossible. If DARWIN discovers dark matter before hitting the neutrino background, says Baudis, that might justify the construction of an even larger, 100-ton instrument to study the stuff in more detail. "But that would really be the end of the road. If there's no such discovery, DARWIN is going to be the last of its kind."

Of course, the solution to the dark matter puzzle may come from different types of detectors, focusing not on WIMPs but on other candidate particles. Future large neutrino experiments, like Hyper-Kamiokande in Japan, the Jiangmen Underground Neutrino Observatory in China, and the Deep Underground Neutrino Experiment in the former Homestake gold mine in South Dakota may shed more light on the masses of the three flavors of neutrinos and on the possible existence of an invisible, warm dark-matter sea of much heavier, sterile neutrinos.[9] And while the ADMX experiment in Seattle, described in chapter 23, hasn't turned up axion-like particles yet, scientists are preparing a much larger and more sensitive International Axion Observatory. The observatory's main task will be to search for axions from the Sun, but the facility might also discover dark matter axions in the halo of the Milky Way.[10]

After my visit to the Euclid clean room, I'm enjoying a glass of wine on one of the terraces at the Place du Capitole, surrounded by Toulouse's characteristic red-pink brick buildings. Again, I try to visualize how millions of dark matter particles are streaming through every square centimeter of my body each and every second. Unnoticed, invisible, mysterious. An omnipresent and ghostly substance, permeating our planet, our solar system, the Milky Way galaxy, and every corner of our expanding universe. And scientists have no clue as to the true identity of the stuff.

Within a few generations, we have discovered humanity's diminutive place in space and time. We mapped the interior of the Earth, we solved the secret of the Sun's energy source, and we peered inside the atomic nucleus. We mastered the genetic code of our DNA, and we learned about contagious viruses and how to combat them. We've created artificial intelligence. And yet, despite decades of dedicated effort, the most brilliant minds on the planet have not succeeded in answering one of the most fundamental questions we've ever formulated: What is the material universe made of?

Maybe we're chasing a chimera, as Mordehai Milgrom and Erik Verlinde believe—particle dark matter may not exist at all. It's quite possible that we're not following the right approach; as Sabine Hossenfelder says, we need to step out of "the empty cycle of theoretically inventing new things, build[ing] new detectors to look for them, and then find[ing] nothing, over and over again—a waste of time and money." And who knows, maybe we're just not properly equipped to understand nature at its most profound level. A lizard can never understand thermodynamics, and we don't expect a dog to solve the equations of quantum mechanics—why should *Homo sapiens* be the first animal to fully grasp the workings of the universe? After all, nature is not obliged to be intelligible to our puny 1,300-gram brain.

Still, despite all the setbacks, doubts, null results, and dead ends, scientists are not giving in. If the current experiment doesn't yield a solution, the next one may, or the next. We cannot tell nature how to behave, as Baudis says, "but we must keep our hopes up. There have been many examples where it took decades before theoretical predictions were borne out by observations. As an experimentalist, you can't just live in the here and now—you're always preparing for the next phase."

Particle physicist Suzan Başeğmez of the National Institute for Subatomic Physics Nikhef in Amsterdam is even more outspoken.[11] Born in Turkey two years after Jim Peebles published his pioneering paper on nonbaryonic cold dark matter, Başeğmez moved to CERN in 2007 and has lived and worked in the Netherlands since 2018. "Dark matter is one of the biggest mysteries in modern science," she says, "and that's why we never give up. I really hope the problem will be solved within the next ten years or so. If by then we're still empty-handed, we may need to start thinking of something new. Or design new experiments."

Başeğmez is part of three collaborations that may play a role in solving the puzzle. These experiments are hunting for signs of dark matter on the ground, in the ocean, and high above our heads. At CERN, the CMS detector may one day discover dark matter particles in the debris from proton collisions. In the Mediterranean Sea, European institutes are building the underwater KM3NeT neutrino detector, which may spot neutrinos that are produced when (and if) dark matter particles annihilate or decay.[12] And since its spectacular and expensive repair operation in late 2019 and early 2020, Sam Ting's Alpha Magnetic Spectrometer (AMS) onboard the International Space Station will keep collecting cosmic ray data for years to come, potentially finding evidence for dark matter annihilation in the universe.

"Working on three experiments at the same time makes it easier to combine the different measurements and to correlate the data," Başeğmez says. "For instance, using various theoretical models, the AMS data can tell us what kind of neutrino signal we may expect in KM3NeT." Meanwhile, she remains hopeful that direct detection experiments like XENONnT, LUX-ZEPLIN, and DARWIN may turn up a WIMP-like dark matter particle in the near future. As for the production of dark matter in particle colliders, she looks forward to the Future Circular Collider, a proposed multibillion-dollar facility almost four times larger and seven times more powerful than CERN's Large Hadron Collider. "It will definitely be very useful for dark matter research," she says. "We basically have no idea what to expect, but I'm not the person to give up."

And then there are the inventive scientists who pursue completely new ways of tracking down dark matter—methods that weren't possible before because technology wasn't advanced enough and in some cases still isn't. For instance, building on an old idea developed by Daniel Snowden-Ifft, Eric Freeman, and Bruford Price, some physicists believe fossil traces of dark matter particles may be found in certain subterranean minerals.[13] A WIMP interaction would give an atomic nucleus a small kick, explains Sebastian Baum of Stanford University, and the energetic nucleus would mess up the crystal structure of the mineral, leaving a discernible microscopic track at most a few tens of nanometers in length. "At depths of more than five kilometers, completely shielded from cosmic rays, such telltale tracks will have been accumulating for hundreds of millions of years," says Baum. "It takes time to convince people, and the technique hasn't been demonstrated yet, but 'paleodetectors' could be a reality a decade from now."[14]

Even more challenging is the gravitational-coupling detector that's being developed by dark matter hunter Rafael Lang and his team at Purdue University. Using a matrix of many millions of tiny zeptonewton sensors sensitive to incredibly small forces—a zeptonewton, or 10^{-21} N, is about one ten-millionth of the weight of a typical bacterium—it might be possible to detect the gravitational effect of a passing dark matter particle, provided the culprit is extremely massive, as suggested by some speculative theories. "We all agree that it's a bit crazy," Lang told journalist Adam Mann in 2020, "but I think everybody will give you a different idea of how crazy it is."[15]

"Crazy" may well be the best word to describe the dark matter conundrum. Crazy and maddening. May 2022 marks exactly one century since Jacobus Kapteyn introduced the world to dark matter as we know it, in his landmark *Astrophysical Journal* paper. Since then astronomers have studied stars, galaxies, and clusters in ever-increasing detail. Their quest to figure out the material makeup of the universe has led to the big bang theory, the understanding of nucleosynthesis in the first few minutes of cosmic history, and the discovery of the cosmic microwave background. Researchers measured the rotation of galaxies, compiled 3D maps of the universe, and used supercomputers to simulate the growth of large-scale structure. Gravitational lenses and distant supernova explosions provided new opportunities to study the distribution of matter and the accelerating expansion of empty space.

Meanwhile, particle physicists discovered neutrinos, antimatter, quarks, and force-carrying bosons, none of which were known to

Kapteyn and his contemporaries. These later investigators developed the successful Standard Model and built ever-larger colliders and underground detectors to test its predictions and look for deviations that might hint at the existence of unknown particles. Their experimental explorations have yielded revelations about the subatomic world, and their theoretical endeavors may one day result in a fruitful marriage of general relativity and quantum mechanics—the Holy Grail of fundamental physics.

But the true nature of dark matter is still a mystery. Despite the efforts of many hundreds of persistent scientists, petabytes of data, and thousands of elaborate publications, we still don't know the identity of more than 80 percent of the material universe. We feel the flapping of an ear and the sharpness of a tusk. We hear the stamping of a foot and the snorting of a trunk. In particular, we experience the massive bulk. But we have no clue about the elephant itself.

Maybe that's OK. Maybe the decades-long search for dark matter is the best possible catalyst for the scientific investigation of both the macro and micro cosmos, just like the quest for extraterrestrial life has inspired planetary exploration, astrochemistry, and the hunt for extrasolar planets. Even if we never reach the destination, there are incredible views to enjoy along the road.

Dark matter governs our universe. Without it, we probably wouldn't be here to wonder about the nature of the cosmos. With it, we will never cease to look for answers. One way or the other, it defines who we are.

Notes

1. Matter, but Not as We Know It

1. James Peebles, interview with author, January 17, 2020, Princeton University.

2. James Peebles, interviewed by Martin Harwit, September 27, 1984, Princeton University, Oral History Interviews, American Institute of Physics, https://www.aip.org/history-programs/niels-bohr-library/oral-histories/4814.

3. P. J. E. Peebles, *Physical Cosmology* (Princeton: Princeton University Press, 1971).

4. P. J. E. Peebles, "How Physical Cosmology Grew," Nobel lecture, December 8, 2019, https://www.nobelprize.org/prizes/physics/2019/peebles/lecture.

5. James Peebles, "Nobel Prize in Physics 2019: Official Interview," telephone interview by Adam Smith, December 6, 2019, https://www.nobelprize.org/prizes/physics/2019/peebles/interview.

2. Underground Phantoms

1. Laboratori Nazionali del Gran Sasso (LNGS), https://www.lngs.infn.it/en.

2. Rafael Lang, "The XENON Experiment: Enlightening the Dark," Xenon Dark Matter Project, April 14, 2017, http://www.xenon1t.org.

3. H. G. Wells, *The Invisible Man* (London: C. Arthur Pearson, 1897).

4. I visited l'Aquila and the Laboratori Nazionali del Gran Sasso on November 4 and 5, 2019.

5. Borexino is the Italian diminutive form of BOREX, which stands for BORon solar neutrino EXperiment.

6. CUPID: CUORE Upgrade with Particle Identification, where CUORE stands for Cryogenic Underground Observatory for Rare Events; VIP: VIolation of the Pauli exclusion principle; COBRA is a shorthand for Cadmium Zinc Telluride 0-Neutrino Double-Beta; GERDA: GERmanium Detector Array.

7. CRESST: Cryogenic Rare Event Search with Superconducting Thermometers; DAMA: DArk MAtter experiment (see chapter 19); COSINUS: Cryogenic Observatory for SIgnatures seen in Next-generation Underground Searches.

3. The Pioneers

1. A concise historical overview of dark matter can be found in G. Bertone and D. Hooper, "History of Dark Matter," *Reviews of Modern Physics* 90, no. 4 (2018), doi: 10.1103/RevModPhys.90.045002.

2. A nontechnical biography of Kapteyn is P. C. van der Kruit, *Pioneer of Galactic Astronomy: A Biography of Jacobus C. Kapteyn* (New York: Springer, 2021).

3. J. C. Kapteyn, "First Attempt at a Theory of the Arrangement and Motion of the Sidereal System," *The Astrophysical Journal* 55 (1922): 302, doi: 10.1086/142670. The term *matière obscure,* French for dark matter, had already been used in 1906 by Henri Poincaré, who sought to discredit the notion that dark matter comprised a significant portion of the universe.

4. W. Thomson, *Baltimore Lectures on Molecular Dynamics and the Wave Theory of Light* (London: C. J. Clay and Sons, 1904), https://archive.org/details/baltimorelecture00kelviala.

5. Kapteyn, "First Attempt at a Theory of the Arrangement and Motion of the Sidereal System," 302. Italics in original.

6. A nontechnical biography of Oort is P. C. van der Kruit, *Master of Galactic Astronomy: A Biography of Jan Hendrik Oort* (New York: Springer, 2021).

7. J. H. Oort, "The Stars of High Velocity" (PhD diss., University of Groningen, 1926).

8. J. H. Oort, "The Force Exerted by the Stellar System in the Direction Perpendicular to the Galactic Plane and Some Related Problems," *Bulletin of the Astronomical Institutes of the Netherlands* 6, no. 238 (1932): 249–287, https://openaccess.leidenuniv.nl/handle/1887/6025.

9. A recent biography of Zwicky is J. Johnson Jr., *Zwicky: The Outcast Genius Who Unmasked the Universe* (Cambridge, MA: Harvard University Press, 2019).

10. F. Zwicky, "Der Rotverschiebung von extragalaktischen Neblen," *Helvetica Physica Acta* 6, no. 2 (1933): 110–127. English translation available at https://ned.ipac.caltech.edu/level5/March17/Zwicky/translation.pdf.

11. S. Smith, "The Mass of the Virgo Cluster," *The Astrophysical Journal* 83 (1936): 23, doi: 10.1086/143697; F. Zwicky, "On the Masses of Nebulae and Clusters of Nebulae," *The Astrophysical Journal* 86 (1937): 217, doi: 10.1086/143864.

12. F. Zwicky, *Morphological Astronomy* (Berlin: Springer-Verlag, 1957).

13. J. H. Oort, "Note on the Determination of K_z and on the Mass Density Near the Sun," *Bulletin of the Astronomical Institutes of the Netherlands* 494 (1960): 45–53, http://adsabs.harvard.edu/pdf/1960BAN. . . . 15 . . . 45O.

14. Koen Kuijken, telephone interview with author, April 24, 2020.

15. K. Kuijken and G. Gilmore, "The Mass Distribution in the Galactic Disc," *Monthly Notices of the Royal Astronomical Society* 239, no. 3 (1989); part 1: 571–603, doi: 10.1093/mnras/239.2.571; part 2: 605–649, doi: 10.1093/mnras/239.2.605; part 3: 651–654, doi: 10.1093/mnras/239.2.651.

16. G. Schilling, "Altijd Geboeid door de Grootste Structuren in het Heelal," *Zenit* 14 (1987): 358.

4. The Halo Effect

1. The full text of Alicia Suskin Ostriker's "Dark Matter and Dark Energy," can be found at https://poets.org/poem/dark-matter-and-dark-energy.

2. Jeremiah Ostriker, interview with author, January 17, 2020, Columbia University.

3. J. P. Ostriker and J. W.-K. Mark, "Rapidly Rotating Stars. I. The Self-Consistent-Field Method," *The Astrophysical Journal* 151 (1968): 1075–1088, doi: 10.1086/149506. Links to parts 2–8 of the series can be found at https://ui.adsabs.harvard.edu/abs/1968ApJ . . . 151.1075O/abstract.

4. R. H. Miller, K. H. Prendergast, and W. J. Quirk, "Numerical Experiments on Spiral Structure," *The Astrophysical Journal* 161 (1970): 903–916, doi: 10.1086/150593; and F. Hohl, "Numerical Experiments with a Disk of Stars," *The Astrophysical Journal* 168 (1971): 343–351, doi: 10.1086/151091.

5. J. P. Ostriker and P. J. E. Peebles, "A Numerical Study of the Stability of Flattened Galaxies: Or, Can Cold Galaxies Survive?" *The Astrophysical Journal* 186 (1973): 467–480, doi: 10.1086/152513.

6. J. H. Oort, in *Transactions of the International Astronomical Union* XIIA (1965), 789.

7. J. P. Ostriker, P. J. E. Peebles, and A. Yahil, "The Size and Mass of Galaxies, and the Mass of the Universe," *The Astrophysical Journal* 193 (1974): L1–L4, doi: 10.1086/181617.

8. F. D. Kahn and L. Woltjer, "Intergalactic Matter and the Galaxy," *The Astrophysical Journal* 130, no. 3 (1959): 705–717, doi: 10.1086/146762.

9. J. Einasto, A. Kaasik, and E. Saar, "Dynamic Evidence on Massive Coronas of Galaxies," *Nature* 250 (1974): 309–310, doi: 10.1038/250309a0.

10. J. P. Ostriker and S. Mitton, *Heart of Darkness: Unraveling the Mysteries of the Invisible Universe* (Princeton: Princeton University Press, 2013).

5. Flattening the Curve

1. Kent Ford, interview with author, January 13, 2020, Millboro Springs, VA.

2. V. C. Rubin and W. K. Ford, Jr., "Rotation of the Andromeda Nebula from a Spectroscopic Survey of Emission Regions," *The Astrophysical Journal* 159 (1970): 379–403, doi: 10.1086/150317.

3. V. C. Rubin, W. K. Ford, Jr,. and N. Thonnard, "Extended Rotation Curves of High-Luminosity Spiral Galaxies. IV. Systematic Dynamical Properties, Sa→Sc," *The Astrophysical Journal* 225 (1978): L107–L113, doi: 10.1086/182804.

4. V. C. Rubin, W. K. Ford, Jr., and N. Thonnard, "Rotational Properties of 21 SC Galaxies with a Large Range of Luminosities and Radii, from NGC 4605 (R = 4kpc) to UGC 2885 (R = 122kpc)," *The Astrophysical Journal* 238 (1980): 471–487, doi: 10.1086/158003.

5. W. Tucker and K. Tucker, *The Dark Matter: Contemporary Science's Quest for the Mass Hidden in Our Universe* (New York: William Morrow, 1988).

6. L. Randall, "Why Vera Rubin Deserved a Nobel," *New York Times,* January 4, 2017.

6. Cosmic Cartography

1. Vera C. Rubin Observatory, https://www.lsst.org.

2. My June 2019 visit to Chile was sponsored by SNP Natuurreizen.

3. I visited Cerro Pachón on June 26, 2019.

4. M. Seldner, B. Siebers, E. J. Groth, and P. J. E. Peebles, "New Reduction of the Lick Catalog of Galaxies," *The Astronomical Journal* 82 (1977): 249–256, plates 313–314, doi: 10.1086/112039.

5. M. Davis, J. Huchra, D. W. Latham, and J. Tonry, "A Survey of Galaxy Redshifts. II. The Large Scale Space Distribution," *The Astrophysical Journal* 253 (1982): 423–445, doi: 10.1086/159646. Links to the other papers in the series can be found at https://ui.adsabs.harvard.edu/abs/1979AJ. . . . .84.1511T/abstract.

6. John Huchra, "The CfA Redshift Survey," n.d., Center for Astrophysics, Cambridge, MA, https://www.cfa.harvard.edu/~dfabricant/huchra/zcat.

7. Matthew Colless, "The 2dF Galaxy Redshift Survey," final data release, June 30, 2003, http://www.2dfgrs.net.

8. The Sloan Digital Sky Survey, https://www.sdss.org.

7. Big Bang Baryons

1. C. H. Payne, "Stellar Atmospheres: A Contribution to the Observational Study of High Temperature in the Reversing Layers of Stars" (PhD diss., University of Cambridge, 1925).

2. A. S. Eddington, "The Internal Constitution of the Stars," *Nature* 106, no. 2653 (1920): 14–20, doi: 10.1038/106014a0.

3. R. A. Alpher, H. Bethe, and G. Gamow, "The Origin of Chemical Elements," *Physical Review* 73, no. 7 (1948): 803–804, doi: 10.1103/PhysRev.73.803.

4. F. Hoyle, "The Synthesis of the Elements from Hydrogen," *Monthly Notices of the Royal Astronomical Society* 106, no. 5 (1946): 343–383, doi: 10.1093/mnras/106.5.343; F. Hoyle, "On Nuclear Reactions Occurring in Very Hot Stars. I. The Synthesis of Elements from Carbon to Nickel," *Astrophysical Journal Supplement* 1 (1954): 121–146, doi: 10.1086/190005; E. Burbidge, G. R. Burbidge, W. A. Fowler, and F. Hoyle, "Synthesis of the Elements in Stars," *Reviews of Modern Physics* 29 (1957): 547–650, doi: 10.1103/RevModPhys.29.547.

5. P. J. E. Peebles, "Primordial Helium Abundance and the Primordial Fireball," *The Astrophysical Journal* 146 (1966): 542–552, doi: 10.1086/148918.

6. R. V. Wagoner, W. A. Fowler, and F. Hoyle, "On the Synthesis of Elements at Very High Temperatures," *The Astrophysical Journal* 148 (1967): 3–49, doi: 10.1086/149126.

7. R. V. Wagoner, "Big-Bang Nucleosynthesis Revisited," *The Astrophysical Journal* 179 (1973): 343–360, doi: 10.1086/151873.

8. J. B. Rogerson and D. G. York, "Interstellar Deuterium Abundance in the Direction of Beta Centauri," *The Astrophysical Journal* 186 (1973): L95, doi: 10.1086/181366.

9. D. G. York and J. B. Rogerson, "The Abundance of Deuterium Relative to Hydrogen in Interstellar Space," *The Astrophysical Journal* 203 (1976): 378–385, doi: 10.1086/154089.

10. J. R. Gott III, J. E. Gunn, S. N. Schramm, and B. M. Tinsley, "An Unbound Universe?" *The Astrophysical Journal* 194 (1974): 543–553, doi: 10.1086/153273.

8. Radio Recollections

1. Albert Bosma, interview with author, November 11, 2019, Westerbork radio observatory.

2. K. Freeman and G. McNamara, *In Search of Dark Matter* (New York: Springer, 2006).

3. H. W. Babcock, "The Rotation of the Andromeda Nebula," *Lick Observatory Bulletin* 19, no. 498 (1939): 41–51, doi: 10.5479/ADS/bib/1939LicOB.19.41B.

4. W. Baade and N. U. Mayall, "Distribution and Motions of Gaseous Masses in Spirals," in *Problems of Cosmical Aerodynamics; Proceedings of a Symposium on the Motion of Gaseous Masses of Cosmical Dimensions* (Paris: IAU and IUTAP, 1951).

5. K. C. Freeman, "On the Disks of Spiral and So Galaxies," *The Astrophysical Journal* 160 (1970): 811–830, doi: 10.1086/150474.

6. National Radio Astronomy Observatory, https://public.nrao.edu.

7. H. I. Ewen and E. M. Purcell, "Observation of a Line in the Galactic Radio Spectrum: Radiation from Galactic Hydrogen at 1,420 Mc./sec.," *Nature* 168 (1951): 356, doi: 10.1038/168356a0; C. A. Muller and J. H. Oort, "Observation of a Line in the Galactic Radio Spectrum: The Interstellar Hydrogen Line at 1,420 Mc./sec., and an Estimate of Galactic Rotation," *Nature* 368 (1951): 357–358, doi: 10.1038/168357a0; J. L. Pawsey, *Nature* 368 (1951): 358, doi: 10.1038/168358a0.

8. H. C. van de Hulst, E. Raimond, and H. van Woerden, "Rotation and Density Distribution of the Andromeda Nebula Derived from Observations of

the 21-cm Line," *Bulletin of the Astronomical Institutes of the Netherlands* 14, no. 480 (1957): 1–16, https://openaccess.leidenuniv.nl/handle/1887/5894.

9. Hugo van Woerden, telephone interview with author, March 31, 2020. Hugo van Woerden passed away on September 4, 2020, at the age of ninety-four.

10. SETI Institute, https://www.seti.org.

11. Seth Shostak, interview with author via Zoom, June 16, 2020.

12. G. Cocconi and P. Morrison, "Searching for Interstellar Communications," *Nature* 184 (1959): 844–846, doi: 10.1038/184844a0.

13. D. H. Rogstad and G. S. Shostak, "Gross Properties of Five Scd Galaxies as Determined from 21-centimeter Observations," *The Astrophysical Journal* 176 (1972): 315–321, doi: 10.1086/151636.

14. M. S. Roberts, "A High-Resolution 21-cm Hydrogen-Line Survey of the Andromeda Nebula," *The Astrophysical Journal* 144 (1966): 639–656, doi: 10.1086/148645.

15. Morton Roberts, interview with author via Zoom, June 17, 2020.

16. M. S. Roberts and R. N. Whitehurst, "The Rotation Curve and Geometry of M31 at Large Galactocentric Distances," *The Astrophysical Journal* 201 (1975): 327–346, doi: 10.1086/153889.

17. Westerbork Synthesis Radio Telescope, https://www.astron.nl/telescopes /wsrt-apertif.

18. A. Bosma, "The Distribution and Kinematics of Neutral Hydrogen in Spiral Galaxies of Various Morphological Types" (PhD diss., University of Groningen, 1978).

19. Katherine Freese, *The Cosmic Cocktail: Three Parts Dark Matter* (Princeton: Princeton University Press, 2016).

20. N. A. Bahcall, "Vera Rubin (1928–2016)," *Nature* 542 (2017): 32, doi: 10.1038/542032a.

21. A. Bosma, "Vera Rubin and the Dark Matter Problem," *Nature* 543 (2017): 179, doi: 10.1038/543179d.

9. Into the Cold

1. S. M. Faber and J. Gallagher, "Masses and Mass-to-Light Ratios of Galaxies," *Annual Review of Astronomy and Astrophysics* 17 (1979): 135–187, doi: 10.1146/annurev.aa.17.090179.001031.

2. Sandra Faber, telephone interview with author, March 28, 2020.

3. Jaan Einasto, "Dark Matter: Early Considerations," in *Frontiers of Cosmology*, ed. A. Blanchard and M. Signore, NATO Science Series (Dordrecht: Springer, 2005).

4. Joel Primack, interview with author via Zoom, March 24, 2020.

5. G. R. Blumenthal, H. Pagels, and J. R. Primack, "Galaxy Formation by Dissipationless Particles Heavier than Neutrinos," *Nature* 299 (1982): 37–38, doi: 10.1038/299037a0.

6. P. J. E. Peebles, "Large-Scale Background Temperature and Mass Fluctuations Due to Scale-Invariant Primeval Perturbations," *The Astrophysical Journal Letters* 263 (1982): L1, doi: 10.1086/183911.

7. The use of the words "cold" and "hot" for slow-moving and fast-moving (relativistic) particles, respectively, was introduced by Joel Primack and Dick Bond in 1983.

8. G. R. Blumenthal, S. M. Faber, J. R. Primack, and M. J. Rees, "Formation of Galaxies and Large-Scale Structure with Cold Dark Matter," *Nature* 311 (1984): 517–525, doi: 10.1038/311517a0.

10. Miraculous WIMPs

1. I visited CERN in June 2019, during a group visit organized by the Dutch edition of *New Scientist* magazine.

2. Large Hadron Collider, https://home.cern/science/accelerators/large-hadron-collider.

3. The name "WIMP" was coined in Gary Steigman and Michael Turner, "Cosmological Constraints on the Properties of Weakly Interacting Massive Particles," *Nuclear Physics B* 253 (1985): 375–386, doi: 10.1016/0550-3213(85)90537-1.

4. J.-L. Gervais and B. Sakita, "Field Theory Interpretation of Supergauges in Dual Models," *Nuclear Physics B* 34 (1971): 632–639, doi: 10.1016/0550-3213(71)90351-8; Y. A. Gol'fand and E. P. Likhtman, "Extension of the Algebra of Poincare Group Generators and Violation of p Invariance," *JETP Letters* 13 (1971): 323, doi: 10.1142/9789814542340_0001; D. V. Kolkov and V. P. Akulov, *Prisma Zh. Eksp. Teor. Fiz.* 16 (1972): 621; J. Wess and B. Zumino, "Supergauge Transformations in Four Dimensions," *Nuclear Physics B* 70 (1974): 39–50, doi: 10.1016/0550-3213(74)90355-1.

5. A Toroidal LHC ApparatuS, https://atlas.cern.

6. John Ellis, interview with author, June 6, 2019, at CERN.

7. J. R. Ellis, J. S. Hagelin, D. V. Nanopoulos, K. A. Olive, and M. Srednicki, "Supersymmetric Relics from the Big Bang," *Nuclear Physics B* 238 (1984): 453–476, doi: 10.1016/0550-3213(84)90461-9.

11. Simulating the Universe

1. IllustrisTNG Project, https://www.tng-project.org.

2. George Efstathiou, Carlos Frenk, and Simon White, interviews with author, September 16, 2019, during the tenth-anniversary symposium of the Kavli Institute for Cosmology in Cambridge, UK.

3. W. H. Press and P. Schechter, "Formation of Galaxies and Clusters of Galaxies by Self-Similar Gravitational Condensation," *The Astrophysical Journal* 187 (1974): 425–438, doi: 10.1086/152650.

4. S. D. M. White, C. S. Frenk, and M. Davis, "Clustering in a Neutrino-Dominated Universe," *The Astrophysical Journal* 274 (1983): L1–L5, doi: 1086/184139.

5. M. Davis, G. Efstathiou, C. S. Frenk, and S. D. M. White, "The Evolution of Large-Scale Structure in a Universe Dominated by Cold Dark Matter," *The Astrophysical Journal* 292 (1985): 371–394, doi: 10.1086/163168.

6. C. S. Frenk, S. D. M. White, G. Efstathiou, and M. Davis, "Cold Dark Matter, the Structure of Galactic Haloes and the Origin of the Hubble Sequence," *Nature* 317 (1985): 595–597, doi: 10.1038/317595a0.

7. S. D. M. White, C. S. Frenk, M. Davis, and G. Efstathiou, "Clusters, Filaments, and Voids in a Universe Dominated by Cold Dark Matter," *The Astrophysical Journal* 313 (1987): 505–516, doi: 10.1086/164990; S. D. M. White, M. Davis, G. Efstathiou, and C. S. Frenk, "Galaxy Distribution in a Cold Dark Matter Universe," *Nature* 330 (1987): 451–453, doi: 10.1038/330451a0; C. S. Frenk, S. D. M. White, M. Davis, and G. Efstathiou, "The Formation of Dark Halos in a Universe Dominated by Cold Dark Matter," *The Astrophysical Journal* 327 (1988): 507–525, doi: 10.1086/166213.

8. S. D. M. White, J. F. Navarro, A. E. Evrard, and C. S. Frenk, "The Baryon Content of Galaxy Clusters: A Challenge to Cosmological Orthodoxy," *Nature* 366 (1993): 429–433, doi: 10.1038/366429a0.

9. Millennium Simulation Project, https://wwwmpa.mpa-garching.mpg.de/galform/virgo/millennium.

10. V. Springel, S. D. M. White, A. Jenkins, et al., "Simulations of the Formation, Evolution and Clustering of Galaxies and Quasars," *Nature* 435 (2005): 629–636, doi: 10.1038/nature03597.

11. EAGLE simulations, http://eagle.strw.leidenuniv.nl.

12. J. Schaye, R. A. Crain, R. G. Bower, et al., "The EAGLE Project: Simulating the Evolution and Assembly of Galaxies and Their Environments," *Monthly Notices of the Royal Astronomical Society* 446 (2015): 521–554, doi: 10.1093/mnras /stu2058; "A Simulation of the Universe with Realistic Galaxies," Durham University press release, January 2, 2015, https://www.dur.ac.uk/news/newsitem /?itemno=23257.

12. The Heretics

1. W. Tucker and K. Tucker, *The Dark Matter: Contemporary Science's Quest for the Mass Hidden in Our Universe* (New York: William Morrow, 1988).

2. T. Standage, *The Neptune File: Planet Detectives and the Discovery of Worlds Unseen* (London: Penguin, 2000).

3. T. Levenson, *The Hunt for Vulcan . . . and How Albert Einstein Destroyed a Planet, Discovered Relativity, and Deciphered the Universe* (New York: Random House, 2015).

4. A. Finzi, "On the Validity of Newton's Law at a Long Distance," *Monthly Notices of the Royal Astronomical Society* 127 (1963): 21–30, doi: 10.1093/mnras /127.1.21.

5. Mordehai Milgrom, interview with author, September 23, 2019, at the workshop The Functioning of Galaxies: Challenges for Newtonian and Milgromian Dynamics, Bonn, Germany.

6. M. Milgrom, "A Modification of the Newtonian Dynamics as a Possible Alternative to the Hidden Mass Hypothesis," *The Astrophysical Journal* 270 (1983): 365–370, doi: 10.1086/161130; M. Milgrom, "A Modification of the Newtonian Dynamics: Implications for Galaxies," *The Astrophysical Journal* 270 (1983): 371–383, doi: 10.168/161131; M. Milgrom, "A Modification of the Newtonian Dynamics: Implications for Galaxy Systems," *The Astrophysical Journal* 270 (1983): 384–389, doi: 10.1086/161132.

7. J. D. Bekenstein, "Relativistic Gravitation Theory for the Modified Newtonian Dynamics Paradigm," *Physical Review D* 70 (2004), art. 083509, doi: 10.1103/PhysRevD.70.083509.

8. R. H. Sandes, "Does GW170817 Falsify MOND?" *International Journal of Modern Physics D* 27 (2018), doi: 10.1142/S0218271818470272.

9. C. Skordis and T. Złośnik, "Gravitational Alternatives to Dark Matter with Tensor Mode Speed Equaling the Speed of Light," *Physical Review D* 100 (2019), art.104013, doi: 10.1103/PhysRevD.100.104013; C. Skordis and T. Złośnik, "New Relativistic Theory for Modified Newtonian Dynamics," *Physical Review Letters* 127, no. 16 (2021), doi: 10.1103/PhysRevLett.127.161302.

10. Stacy McGaugh, telephone interview with author, March 30, 2020.

11. S. McGaugh, "Predictions and Outcomes for the Dynamics of Rotating Galaxies," *Galaxies* 8, no. 2 (2020): 35, doi: 10.3390/galaxies8020035.

12. G. Schilling, "Battlefield Galactica: Dark Matter vs. MOND," *Sky & Telescope* 113, no. 4 (April 2007): 30–36.

13. R. H. Sanders, *The Dark Matter Problem: A Historical Perspective* (Cambridge: Cambridge University Press, 2010).

13. Behind the Lens

1. A. Einstein, "Lens-like Action of a Star by the Deviation of Light in the Gravitational Field," *Science* 84 (1936): 506–507, doi: 10.1126/science.84.2188.506.

2. F. Zwicky, "Nebulae as Gravitational Lenses," *Physical Review* 51 (1937): 290, doi: 10.1103/PhysRev.51.290.

3. D. Walsh, R. F. Carswell, and R. J. Weymann, "0957 + 561 A, B: Twin Quasistellar Objects or Gravitational Lens?" *Nature* 279 (1979): 381–384, doi: 10.1038/279381a0.

4. J. E. Gunn, "On the Propagation of Light in Inhomogeneous Cosmologies. I. Mean Effects," *The Astrophysical Journal* 150 (1967): 737–753, doi: 10.1086/149378.

5. J. A. Tyson, F. Valdes, J. F. Jarvis, and A. P. Millis Jr., "Galaxy Mass Distribution from Gravitational Light Deflection," *The Astrophysical Journal* 281 (1984): L59–L62, doi: 10.1086/184285.

6. M. Markevitch, A. H. Gonzalez, L. David, et al., "A Textbook Example of a Bow Shock in the Merging Galaxy Cluster 1E 0657–56," *The Astrophysical Journal Letters* 567 (2002): L27–L31, doi: 10.1086/339619.

7. Douglas Clowe, interview with author via Zoom, July 21, 2020.

8. D. Clowe, A. Gonzalez, and M. Markevitch, "Weak-Lensing Mass Reconstruction of the Interacting Cluster 1E 0657-558: Direct Evidence for the Existence of Dark Matter," *The Astrophysical Journal* 604 (2004): 596–603, doi: 10.1086/381970.

9. D. Clowe, M. Bradač, A. H. Gonzalez, et al., "A Direct Empirical Proof of the Existence of Dark Matter," *The Astrophysical Journal* 648 (2006): L109–L113, doi: 10.1086/508162.

10. "NASA Finds Direct Proof of Dark Matter," NASA press release 06-297, August 21, 2006, https://chandra.harvard.edu/press/06_releases/press_082106.html.

11. R. H. Sanders, *The Dark Matter Problem: A Historical Perspective* (Cambridge: Cambridge University Press, 2010).

12. R. Massey, T. Kitching, and J. Richard, "The Dark Matter of Gravitational Lensing," *Reports on Progress in Physics* 73, no. 8 (2010), 086901, doi: 10.1088/0034-4885/73/8/086901.

13. Hubble Frontier Fields Program, https://frontierfields.org.

14. L. van Waerbeke, Y. Mellier, T. Erben, et al., "Detection of Correlated Galaxy Ellipticities from CFHT Data: First Evidence for Gravitational Lensing by Large-Scale Structures," *Astronomy and Astrophysics* 358 (2000): 30–44, https://arxiv.org/abs/astro-ph/0002500; D. M. Wittman, J. A. Tyson, G. Bernstein, et al., "Detection of Weak Gravitational Lensing Distortions of Distant Galaxies by Cosmic Dark Matter at Large Scales," *Nature* 405 (2000): 143–148, doi: 10.1038/35012001; D. J. Bacon, A. R. Refregier, and R. S. Ellis, "Detection of Weak Gravitational Lensing by Large-Scale Structure," *Monthly Notices of the Royal Astronomical Society* 318 (2000): 625–640, doi: 10.1046/j.1365-8711.2000.03851.x; N. Kaiser, "A New Shear Estimator for Weak-Lensing Observations," *The Astrophysical Journal* 537 (2000): 555–577, doi: 10.1086/309041.

15. Extremely Large Telescope, https://elt.eso.org.

14. MACHO Culture

1. S. Refsdal, "The Gravitational Lens Effect," *Monthly Notices of the Royal Astronomical Society* 128 (1964): 295–306, doi: 10.1093/mnras/128.4.295.

2. A detailed history of gravitational lensing, including more information on Maria Petrou, can be found in David Valls-Gabaud, "Gravitational Lensing:

The Early History," slide deck from presentation of September 17, 2012, http://www.cpt.univ-mrs.fr/~cosmo/EcoleCosmologie/DossierCours11 /Se%CC%81minaires/valls-gabaud.pdf.

3. B. Paczyński, "Gravitational Microlensing by the Galactic Halo," *The Astrophysical Journal* 304 (1986): 1–5, doi: 10.1086/164140.

4. Charles Alcock, telephone interview with author, July 9, 2020.

5. Éric Aubourg, telephone interview with author, July 9, 2020.

6. Optical Gravitational Lensing Experiment, http://ogle.astrouw.edu.pl.

7. C. Alcock, C. W. Akerlof, R. A. Allsman, et al., "Possible Gravitational Microlensing of a Star in the Large Magellanic Cloud," *Nature* 365 (1993): 621–623, doi: 10.1038/365621a0; E. Aubourg, P. Bareyre, S. Bréhin, et al., "Evidence for Gravitational Microlensing by Dark Objects in the Galactic Halo," *Nature* 365 (1993): 623–625, doi: 10.1038/365623a0.

8. C. Hogan, "In Search of the Halo Grail," *Nature* 365 (1993): 602–603, doi: 10.1038/365602a0.

9. C. Alcock, R. A. Allsman, D. Alves, et al., "EROS and MACHO Combined Limits on Planetary-Mass Dark Matter in the Galactic Halo," *The Astrophysical Journal* 499 (1998): L9–L12, doi: 10.1086/311355.

10. C. Alcock, R. A. Allsman, D. Alves, et al., "The MACHO Project: Microlensing Results from 5.7 Years of Large Magellanic Cloud Observations," *The Astrophysical Journal* 542 (2000): 281–307, doi: 10.1086/309512.

11. P. Tisserand, L. Le Guillou, C. Afonso, et al., "Limits on the MACHO Content of the Galactic Halo from the EROS-2 Survey of the Magellanic Clouds," *Astronomy and Astrophysics* 469 (2007): 387–404, doi: 10.1051/0004 -6361:20066017.

15. The Runaway Universe

1. A. R. Sandage, "Cosmology: A Search for Two Numbers," *Physics Today* 23 (1970): 34–41, doi: 10.1063/1.3021960.

2. W. L. Freedman, B. F. Madore, B. K. Gibson, et al., "Final Results from the Hubble Space Telescope Key Project to Measure the Hubble Constant," *The Astrophysical Journal* 533 (2001): 47–72, doi: 10.1086/320638.

3. A. H. Guth, "Inflationary Universe: A Possible Solution to the Horizon and Flatness Problems," *Physical Review D* 23 (1981): 347–356, doi: 10.1103/Phys RevD.23.347.

4. A. H. Guth, *The Inflationary Universe: The Quest for a New Theory of Cosmic Origins* (New York: Basic Books, 1998).

5. A. G. Riess, A. V. Filippenko, P. Challis, et al., "Observational Evidence from Supernovae for an Accelerating Universe and a Cosmological Constant," *The Astronomical Journal* 116 (1998): 1009–1038, doi: 10.1086/300499.

6. S. Perlmutter, G. Aldering, G. Goldhaber, et al., "Measurements of Ω and Λ from 42 High-Redshift Supernovae," *The Astrophysical Journal* 517 (1999): 565–586, doi: 10.1086/307221.

7. D. Overbye, "Studies of Universe's Expansion Win Physics Nobel," *New York Times,* October 4, 2011.

16. Pie in the Sky

1. C. O'Raifeartaigh, "Investigating the Legend of Einstein's 'Biggest Blunder,'" *Physics Today,* October 30, 2018, doi: 10.1063/PT.6.3.20181030a.

2. J. P. Ostriker and P. J. Steinhardt, "The Observational Case for a Low-Density Universe with a Non-Zero Cosmological Constant," *Nature* 377 (1995): 600–602, doi: 10.1038/377600a0.

3. R. Panek, *The 4% Universe: Dark Matter, Dark Energy, and the Race to Discover the Rest of Reality* (New York: Houghton Mifflin Harcourt, 2011).

4. J. Colin, R. Mohayaee, M. Rameez, and S. Sarkar, "Evidence for Anisotropy of Cosmic Acceleration," *Astronomy and Astrophysics* 631 (2019): L13, doi: 10.1051/0004-6361/201936373.

5. Y. Kang Y.-W. Lee, Y.-L. Kim, et al., "Early-Type Host Galaxies of Type Ia Supernovae. II. Evidence for Luminosity Evolution in Supernova Cosmology," *The Astrophysical Journal* 889 (2020): 8, doi: 10.3847/1538-4357/ab5afc.

6. J. Cartwright, "Dark Energy Is the Biggest Mystery in Cosmology, but It May Not Exist at All," *Horizon,* September 3, 2018, https://ec.europa.eu/research-and-innovation/en/horizon-magazine/dark-energy-biggest-mystery-cosmology-it-may-not-exist-all-leading-physicist.

7. "New evidence shows that the key assumption made in the discovery of dark energy is in error," Yonsei University Press Release, January 5, 2020, https://devcms.yonsei.ac.kr/galaxy_en/galaxy01/research.do?mode=view&articleNo=78249.

8. R. R. Caldwell, M. Kamionkowski, and N. N. Weinberg, "Phantom Energy: Dark Energy with $w < -1$ Causes a Cosmic Doomsday," *Physical Review Letters* 91 (2003), 071301, doi: 10.1103/PhysRevLett.91.071301.

17. Telltale Patterns

1. Planck mission, https://sci.esa.int/web/planck. I attended the *Planck* launch in French Guiana on May 14, 2009, on invitation from the European Space Agency.

2. R. K. Sachs and A. M. Wolfe, "Perturbations of a Cosmological Model and Angular Variations of the Microwave Background," *The Astrophysical Journal* 147 (1967): 73–90, doi: 10.1086/148982.

3. N. Aghanim, Y. Akrami, F. Arroja, et al., "Planck 2018 Results. I. Overview and the Cosmological Legacy of Planck," *Astronomy and Astrophysics* 641 (2018), article A1, doi: 10.1051/0004-6361/201833880. Links to the other papers in the series can be found at https://www.cosmos.esa.int/web/planck /publications.

4. N. Aghanim, Y. Akrami, M. Ashdown, et al., "Planck 2018 Results. VI. Cosmological Parameters," *Astronomy and Astrophysics* 641 (2018), article A6, doi: 10.1051/0004-6361/201833910.

5. S. Cole, W. J. Percival, J. A. Peacock, et al., "The 2dF Galaxy Redshift Survey: Power-Spectrum Analysis of the Final Data Set and Cosmological Implications," *Monthly Notices of the Royal Astronomical Society* 362 (2005): 505–534, doi: 10.1111/j.1365-2966.2005.09318.x; D. J. Eisenstein, I. Zehavi, D. W. Hogg, et al., "Detection of the Baryon Acoustic Peak in the Large-Scale Correlation Function of SDSS Luminous Red Galaxies," *The Astrophysical Journal* 633 (2005): 560–574, doi: 10.1086/466512.

18. The Xenon Wars

1. Elena Aprile, interview with author, January 18, 2020, New York.

2. XENON experiment, http://www.xenon1t.org.

3. Sudbury Neutrino Observatory Laboratory, https://www.snolab.ca.

4. Imaging Cosmic And Rare Underground Signals, http://icarus.lngs .infn.it.

5. J. Angle, E. Aprile, F. Arneodo, et al., "First Results from the XENON10 Dark Matter Experiment at the Gran Sasso National Laboratory," *Physical Review Letters* 100 (2008), 021303, doi: 10.1103/PhysRevLett.100.021303.

6. Richard Gaitskell, interview with author, January 16, 2020, Brown University, Providence, RI.

7. The Cryogenic Dark Matter Search (CDMS) has now evolved into SuperCDMS, https://supercdms.slac.stanford.edu.

8. Sanford Underground Research Facility, https://sanfordlab.org.

9. E. Aprile, M. Alfonsi, K. Arisaka, et al., "Dark Matter Results from 225 Live Days of XENON100 Data," *Physical Review Letters* 109 (2012), 181301, doi: 10.1103/PhysRevLett.109.181301; D. S. Akerib, H. M. Araújo, X. Bai, et al., "First Results from the LUX Dark Matter Experiment at the Sanford Underground Research Facility," *Physical Review Letters* 112 (2014), 091303, doi: 10.1103 /PhysRevLett.112.091303.

10. E. Aprile, J. Aalbers, F. Agostini, et al., "First Dark Matter Search Results from the XENON1T Experiment," *Physical Review Letters* 119 (2017), 181301, doi: 10.1103/PhysRevLett.119.181301.

11. PandaX experiment, https://pandax.sjtu.edu.cn.

12. LUX-ZEPLIN experiment, https://lz.lbl.gov.

19. Catching the Wind

1. Rita Bernabei, email message to author, September 11, 2020.

2. M. W. Goodman and E. Witten, "Detectability of Certain Dark-Matter Candidates," *Physical Review D* 31 (1985): 3059–3063, doi: 10.1103/PhysRevD .31.3059.

3. A. K. Drukier, K. Freese, and D. N. Spergel, "Detecting Cold Dark-Matter Candidates," *Physical Review D* 33 (1986): 3495–3608, doi: 10.1103 /physrevd.33.3495.

4. The Cryogenic Dark Matter Search (CDMS) has now evolved into SuperCDMS, https://supercdms.slac.stanford.edu.

5. Expérience pour Detecter Les WIMPs En Site Souterrain, http:// edelweiss.in2p3.fr.

6. DAMA project, http://people.roma2.infn.it/~dama/web/home.html.

7. R. Bernabei, P. Belli, F. Montecchia, et al., "Searching for WIMPs by the Annual Modulation Signature," *Physics Letters B* 424 (1998): 195–201, doi: 10.1016/S0370-2693(98)00172-5.

8. R. Bernabei, P. Belli, F. Cappella, et al., "Dark Matter Search," *La Rivista del Nuovo Cimento* 26 (2003): 1–73.

9. R. Bernabei, P. Belli, F. Cappella, et al., "Final Model Independent Result of DAMA/LIBRA–phase1," *European Physical Journal C* 73 (2013): 1–11, doi: 10.1140/epjc/s10052-013-2648-7.

10. R. Bernabei, P. Belli, A. Bussolotti, et al., "First Model Independent Results from DAMA/LIBRA–phase2," *Nuclear Physics and Atomic Energy* 19, no. 4 (2018): 307–325, doi: 10.15407/jnpae2018.04.307.

11. Reina Maruyama, telephone interview with author, September 29, 2020.

12. COSINE-100 Collaboration, "An Experiment to Search for Dark-Matter Interactions Using Sodium Iodide Detectors," *Nature* 564 (2018): 83–86, doi: 10.1038/s41586-018-0739-1.

13. J. Amaré, S. Cébrian, D. Cintas, et al., "Annual Modulation Results from Three-Year Exposure of ANAIS-112," *Physical Review D* 103 (2021), 102005, doi: 10.1103/PhysRevD.103.102005.

14. SABRE experiment, https://sabre.lngs.infn.it.

15. K. Freese, *The Cosmic Cocktail: Three Parts Dark Matter* (Princeton: Princeton University Press, 2014).

16. A. K. Drukier, K. Freese, A. Lopez, et al., "New Dark Matter Detectors Using DNA or RNA for Nanometer Tracking," arXiv:1206.6809v2; A. K. Drukier, C. Cantor, M. Chonofsky, et al., "New Class of Biological Detectors for WIMPs," *International Journal of Modern Physics A* 29 (2014), 1443007, doi: 10.1142/S0217751X14430076.

20. Messengers from Outer Space

1. Luca Parmitano, Twitter post, February 3, 2020, http://twitter.com/astro_luca/status/1224315152746602497.

2. Samuel Ting, interview with author via Zoom, September 22, 2020.

3. J. Alcaraz, D. Alvisi, B. Alpat, et al., "Protons in Near Earth Orbit," *Physics Letters B* 472 (2000): 215–226, doi: 10.1016/S0370-2693(99)01427-6; J. Alcaraz,

B. Alpat, G. Ambrosi, et al., "Leptons in Near Earth Orbit," *Physics Letters B* 484 (2000): 10–22, doi: 10.1016/S0370-2693(00)00588-8.

4. AMS-02 project, https://ams02.space.

5. O. Adriani, G. C. Barbarino, G. A., Bazilevskaya, et al., "An Anomalous Positron Abundance in Cosmic Rays with Energies 1.5–100 GeV," *Nature* 458 (2009): 607–609, doi: 10.1038/nature07942.

6. G. Brumfiel, "Physicists Await Dark-Matter Confirmation," *Nature* 454 (2008): 808–809, doi: 10.1038/454808b.

7. Fermi Gamma-ray Space Telescope, https://fermi.gsfc.nasa.gov.

8. Tracy Slatyer, interview with author, January 15, 2020, MIT, Cambridge, MA.

9. G. Dobler, D. P. Finkbeiner, I. Cholis, T. R. Slatyer, and N. Weiner, "The Fermi Haze: A Gamma-Ray Counterpart to the Microwave Haze," *The Astrophysical Journal* 717 (2010): 825–842, doi: 10.1088/0004-637X/717/2 /825.

10. M. Su, T. R. Slatyer, and D. P. Finkbeiner, "Giant Gamma-Ray Bubbles from Fermi-LAT: Active Galactic Nucleus Activity or Bipolar Galactic Wind?" *The Astrophysical Journal* 724 (2010): 1044–1082, doi: 10.1088/0004-637X/724/2 /1044.

11. Dan Hooper, telephone interview with author, March 23, 2020.

12. D. Hooper and T. R. Slatyer, "Two Emission Mechanisms in the Fermi Bubbles: A Possible Signal of Annihilating Dark Matter," *Physics of the Dark Universe* 2 (2013): 118–138, doi: 10.1016/j.dark.2013.06.003.

13. R. K. Leane and T. R. Slatyer, "Revival of the Dark Matter Hypothesis for the Galactic Center Gamma-Ray Excess," *Physical Review Letters* 123 (2019), 241101, doi: 10.1103/PhysRevLett.123.241101.

14. I visited the AMS Payload Operations Control Centre at CERN on June 6, 2019.

15. M. Aguilar, G. Alberti, B. Alpat, et al., "First Result from the Alpha Magnetic Spectrometer on the International Space Station: Precision Measurement of the Positron Fraction in Primary Cosmic Rays of 0.5–350 GeV," *Physical Review Letters* 110 (2013), 141102, doi: 10.1103/PhysRevLett.110.141102; M. Aguilar, L. Ali Cavasonza, G. Ambrosi, et al., "Toward Understanding the Origin of Cosmic-Ray Positrons," *Physical Review Letters* 122 (2019), 041102, doi: 10.1103/PhysRevLett.122.041102.

21. Delinquent Dwarfs

1. Pieter van Dokkum, interview by author, January 7, 2020, at the 235th meeting of the American Astronomical Society, Honolulu.

2. Dragonfly Telephoto Array, https://www.dragonflytelescope.org.

3. MOND backers don't agree, by the way: the odd kinematics of the dwarf galaxy could be due to the proximity of the larger galaxy NGC 1052, with its (Mondian) gravitational influence—the so-called external field effect.

4. H. Shapley, "Two Stellar Systems of a New Kind," *Nature* 142 (1938): 715–716, doi: doi.org/ 10.1038 / 142715b0.

5. J. Kormendy and K. C. Freeman, "Scaling Laws for Dark Matter Halos in Late-Type and Dwarf Spheroidal Galaxies," *Proceedings of the International Astronomical Union,* vol. 220: *Dark Matter in Galaxies* (2004): 377–397, doi: 10.1017/S0074180900183706.

6. J. F. Navarro, C. S. Frenk, and S. D. M. White, "The Structure of Cold Dark Matter Halos," *The Astrophysical Journal* 462 (1996): 563–575, doi: 10.1086 /177173.

7. P. G. van Dokkum, *Dragonflies: Magnificent Creatures of Water, Air, and Land* (New Haven: Yale University Press, 2015).

8. New Mexico Skies Observatories, https://www.nmskies.com.

9. P. G. van Dokkum, R. Abraham, A. Merritt, J. Zhang, M. Geha, and C. Conroy, "Forty-Seven Milky Way–Sized, Extremely Diffuse Galaxies in the Coma Cluster," *The Astrophysical Journal Letters* 798 (2015): L45, doi: 10.1088 /2041-8205/798/2/L45.

10. P. G. van Dokkum, A. J. Romanowsky, R. Abraham, et al., "Spectroscopic Confirmation of the Existence of Large, Diffuse Galaxies in the Coma Cluster," *The Astrophysical Journal Letters* 804 (2015): L26, doi: 10.1088/2041 -8205/804/1/L26.

11. P. G. van Dokkum, R. Abraham, J. Brodie, et al., "A High Stellar Velocity Dispersion and ~100 Globular Clusters for the Ultra-Diffuse Galaxy Dragonfly 44," *The Astrophysical Journal Letters* 828 (2016): L6, doi: 10.3847/2041 -8205/828/1/L6.

12. Stacy McGaugh, telephone interview with author, March 30, 2020.

13. P. G. van Dokkum, S. Danieli, Y. Cohen, et al., "A Galaxy Lacking Dark Matter," *Nature* 555 (2018): 629–632, doi: 10.1038/nature25767.

14. P. G. van Dokkum, S. Danieli, R. Abraham, C. Conroy, and A. Romanowsky, "A Second Galaxy Missing Dark Matter in the NGC 1052 Group," *The Astrophysical Journal Letters* 874 (2019): L5, doi: 10.3847/2041-8213/ab0d92.

15. P. Kroupa, C. Theis, and C. M. Boily, "The Great Disk of Milky-Way Satellites and Cosmological Sub-Structures," *Astronomy and Astrophysics* 431 (2005): 517–521, doi: 10.1051/0004-6361:20041122.

16. R. A. Ibata, G. F. Lewis, A. R. Conn, et al., "A Vast, Thin Plane of Corotating Dwarf Galaxies Orbiting the Andromeda Galaxy," *Nature* 493 (2013): 62–65, doi: 10.1038/nature11717.

17. O. Müller, M. Pawlowski, H. Jerjen, and F. Lelli, "A Whirling Plane of Satellite Galaxies around Centaurus A Challenges Cold Dark Matter Cosmology," *Science* 359 (2018): 534–537, doi: 10.1126/science.aao1858.

18. Marcel Pawlowski, interview with author, September 23, 2019, at the workshop The Functioning of Galaxies: Challenges for Newtonian and Milgromian Dynamics, Bonn, Germany.

19. M. Pawlowski, "The Planes of Satellite Galaxies Problem, Suggested Solutions, and Open Questions," *Modern Physics Letters A* 33 (2018), 1830004, doi: 10.1142/S0217732318300045.

22. Cosmological Tension

1. "The Hubble Constant Controversy: Status, Implications and Solutions," WE-Heraeus-Symposium, November 10, 2018, Berlin, https://www.we-heraeus-stiftung.de/veranstaltungen/tagungen/2018/hubble2018.

2. W. L. Freedman, B. F. Madore, B. K. Gibson, et al., "Final Results from the Hubble Space Telescope Key Project to Measure the Hubble Constant," *The Astrophysical Journal* 553 (2001): 47–72, doi: 10.1086/320638.

3. N. Aghanim, Y. Akrami, M. Ashdown, et al., "Planck 2018 Results. VI. Cosmological Parameters," *Astronomy and Astrophysics* 641 (2020), A6, doi: 10.1051/0004-6361/201833910.

4. A. Riess, S. Casertano, W. Yuan, L. M. Macri, and D. Scolnic, "Large Magellanic Cloud Cepheid Standards Provide a 1% Foundation for the Determination of the Hubble Constant and Stronger Evidence for Physics beyond ΛCDM," *The Astrophysical Journal* 876 (2019): 85, doi: 10.3847/1538-4357/ab1422.

5. K. C. Wong, S. H. Suyu, G. C.-F. Chen, et al., "HoLiCOW XIII. A 2.4% Measurement of H_0 from Lensed Quasars: 5.3σ Tension between Early- and Late-Universe Probes," *Monthly Notices of the Royal Astronomical Society* 498 (2020): 1420–1439, doi: 10.1093/mnras/stz3094.

6. W. L. Freedman, B. F. Madore, T. Hoyt, et al., "Calibration of the Tip of the Red Giant Branch (TRGB)," *The Astrophysical Journal* 891 (2020): 57, doi: 10.3847/1538-4357/ab7339.

7. Natalie Wolchover, "Cosmologists Debate How Fast the Universe Is Expanding," *Quanta Magazine,* August 8, 2019, https://www.quantamagazine.org/cosmologists-debate-how-fast-the-universe-is-expanding-20190808.

8. Kilo-Degree Survey, http://kids.strw.leidenuniv.nl.

9. Dark Energy Survey, https://www.darkenergysurvey.org.

10. Hyper Suprime-Cam Subaru Strategic Program, https://hsc.mtk.nao.ac.jp/ssp.

11. E. Di Valentino: "The Tension Cosmology," presentation at the VII Meeting of Fundamental Cosmology, Madrid, September 9–11, 2019, https://agenda.ciemat.es/event/1126/contributions/2119/attachments/1604/1919/divalentino.pdf.

12. Eleonora Di Valentino, interview with author via Zoom, December 21, 2020.

13. George Efstathiou, email message to author, November 11, 2020.

23. Elusive Ghosts

1. Karlsruhe Tritium Neutrino experiment, https://www.katrin.kit.edu.

2. S. Dodelson and L. M. Widrow, "Sterile Neutrinos as Dark Matter," *Physical Review Letters* 72 (1994): 17–20, doi: 10.1103/PhysRevLett.72.17.

3. Y. Fukuda, T. Hayakawa, E. Ichihara, et al., "Evidence for Oscillation of Atmospheric Neutrinos," *Physical Review Letters* 81 (1998): 1562–1567, doi: 10.1103/PhysRevLett.81.1562; Q. R. Ahmad, R. C. Allen, T. C. Andersen, et al., "Measurement of the Rate of $\nu_e + d \rightarrow p + p + e^-$ Interactions Produced by ^{8}B Solar Neutrinos at the Sudbury Neutrino Observatory," *Physical Review Letters* 87 (2001), 071301, doi: 10.1103/PhysRevLett.87.071301.

4. I visited the Karlsruhe Tritium Neutrino experiment on September 5, 2019.

5. M. Aker, K. Altenmüller, N. Arenz, et al., "Improved Upper Limit on the Neutrino Mass from a Direct Kinematic Method by KATRIN," *Physical Review Letters* 123 (2019), 221802, doi: 10.1103/PhysRevLett.123.221802.

6. R. D. Peccei and H. R. Quinn, "CP Conservation in the Presence of Pseudoparticles," *Physical Review Letters* 38 (1977): 1440–1443, doi: 10.1103/Phys RevLett.38.1440.

7. CERN Axion Solar Telescope, https://home.cern/science/experiments /cast.

8. Any Light Particle Search, https://alps.desy.de.

9. P. Sikivie, "Experimental Tests of the 'Invisible' Axion," *Physical Review Letters* 51 (1983): 1415–1417, doi: 10.1103/PhysRevLett.51.1415.

10. Axion Dark Matter eXperiment, https://depts.washington.edu/admx.

11. E. Aprile, J. Aalbers, F. Agostini, et al., "Excess Electronic Recoil Events in XENON1T," *Physical Review D* 102 (2020), 072004, doi: 10.1103/PhysRevD. 102.072004.

24. Dark Crisis

1. Erik Verlinde, interview with author, December 3, 2020, University of Amsterdam.

2. Kathryn Zurek, interview with author via Zoom, September 17, 2020.

3. Sabine Hossenfelder, interview with author via Zoom, December 21, 2020.

4. B. J. Carr and S. W. Hawking, "Black Holes in the Early Universe," *Monthly Notices of the Royal Astronomical Society* 168 (1974): 399–415, doi: 10.1093/mnras /168.2.399.

5. K. Jedamzik, "Primordial Black Hole Dark Matter and the LIGO/Virgo Observations," *Journal of Cosmology and Astroparticle Physics* 2020 (2020), 022, doi: 10.1088/1475-7516/2020/09/022.

6. L. Hui, J. P. Ostriker, S. Tremaine, and E. Witten, "Ultralight Scalars as Cosmological Dark Matter," *Physical Review D* 95 (2017), 043541, doi: 10.1103 /PhysRevD.95.043541.

7. K. Vattis, S. M. Koushiappas, and A. Loeb, "Dark Matter Decaying in the Late Universe Can Relieve the H_0 Tension," *Physical Review D* 99 (2019), 121302, doi: 10.1103/PhysRevD.99.121302.

8. ForwArd SEaRch experiment, https://faser.web.cern.ch.

9. Jamie Boyd, interview with author, June 4, 2019, CERN.

10. L. Berezhiani and J. Khoury, "Theory of Dark Matter Superfluidity," *Physical Review D* 92 (2015), 103510, doi: 10.1103/PhysRevD.92.103510.

11. S. Hossenfelder, "Superfluid Dark Matter," March 24, 2019, https://youtu.be/468cyBZ_cq4.

12. E. P. Verlinde, "Emergent Gravity and the Dark Universe," *SciPost Physics* 2 (2017), 016, doi: 10.21468/SciPostPhys.2.3.016.

25. Seeing the Invisible

1. I visited Airbus Defence and Space in Toulouse on August 3, 2020.

2. Euclid mission, https://sci.esa.int/web/euclid.

3. Nancy Grace Roman Space Telescope, https://roman.gsfc.nasa.gov.

4. Dark Energy Spectroscopic Instrument, https://www.desi.lbl.gov.

5. G. Bertone and T. M. P. Tait, "A New Era in the Search for Dark Matter," *Nature* 562 (2018): 51–56, doi: 10.1038/s41586-018-0542-z.

6. Laura Baudis, interview with author via Zoom, December 14, 2020.

7. DARWIN project, https://darwin.physik.uzh.ch.

8. E. Gibney, "Last Chance for WIMPs: Physicists Launch All-Out Hunt for Dark-Matter Candidate," *Nature* 586 (2020): 344–345, doi: 10.1038/d41586-020-02741-3.

9. Hyper-Kamiokande, https://www.hyperk.org; Jiangmen Underground Neutrino Observatory, http://juno.ihep.cas.cn; Deep Underground Neutrino Experiment, https://www.dunescience.org.

10. International Axion Observatory, https://iaxo.web.cern.ch.

11. Suzan Başeğmez, telephone interview with author, January 4, 2021.

12. KM3NeT, https://www.km3net.org.

13. D. P. Snowden-Ifft, E. S. Freeman, and P. B. Price, "Limits on Dark Matter Using Ancient Mica," *Physical Review Letters* 74 (1995): 4133–4136, doi: 10.1103/PhysRevLett.74.4133.

14. Sebastian Baum, telephone interview with author, September 30, 2020.

15. A. Mann, "The Detector with a Billion Sensors That May Finally Snare Dark Matter," *New Scientist,* July 1, 2020, https://www.newscientist.com/article/mg24632891-200-the-detector-with-a-billion-sensors-that-may-finally-snare-dark-matter.

Acknowledgments

Many people have helped me to realize this book. First, thanks to my agent, Peter Tallack at the Science Factory, and to Janice Audet at Harvard University Press for their trust, enthusiasm, and support. I am very grateful as well for the comments of the two anonymous reviewers who checked my draft manuscript for factual errors and inconsistencies. Thanks to Simon Waxman for meticulously improving the style and grammar of my original manuscript. And thanks to Avi Loeb for writing the foreword.

Most of all, I want to thank the many astrophysicists, radio astronomers, cosmologists, particle physicists, theorists, computer wizards, and instrument builders who generously showed me around their research facilities, shared their stories and thoughts with me, and helped me to improve the manuscript. Of course, any remaining factual errors are my own fault. In particular, I wish to thank Bob Abraham, Charles Alcock, Elena Aprile, Éric Aubourg, Suzan Başeğmez, Laura Baudis, Sebastian Baum, Melissa van Beekveld, Rita Bernabei, Gianfranco Bertone, Albert Bosma, Jamie Boyd, Laurent Brouard, Douglas Clowe, Dan Coe, Auke Pieter Colijn, Patrick Decowski, Eleonora Di Valentino, Pieter van Dokkum, George Efstathiou, Daniel Eisenstein, John Ellis, Sandra Faber, Kent and Ellen Ford, Katherine Freese, Carlos Frenk, Rick Gaitskell, Amina Helmi, Dan Hooper, Sabine Hossenfelder, Koen Kuijken,

Eric Laenen, Avi Loeb, Jennifer Lotz, Reina Maruyama, Stacy McGaugh, Daan Meerburg, Mordehai Milgrom, Jerry Ostriker, Mercedes Paniccia, Marcel Pawlowski, Jim Peebles, Tristan du Pree, Joel Primack, Morton Roberts, Diederik Roest, Gray Rybka, Joop Schaye, Jacques and Renee Sebag, Seth Shostak, Tracy Slatyer, Markus Steidl, Jaco de Swart, Samuel Ting, Erik Verlinde, Ivo van Vulpen, Simon White, the late Hugo van Woerden, Alfredo Zenteno, and Kathryn Zurek.

Portions of chapter 22 were first published as "Constant Controversy" in the June 2019 issue of *Sky & Telescope*. They are reproduced here with permission.

Image Credits

Page 13: Courtesy P. J. E. Peebles

p. 30: XENON Collaboration

p. 43: Rudolf Riedl

p. 54: ESO / L. Calçada

p. 61: Carnegie Institution for Science / DTM Archives

p. 75: Todd Mason, Mason Productions Inc. / Rubin Observatory / NSF / AURA

p. 91: NASA

p. 103: Marcel Schmeier

p. 116: © The Regents of the University of California. Courtesy Special Collections, University Library, University of California, Santa Cruz: US Santa Cruz Photography Services Photographs.

p. 123: CERN

p. 135: D. Nelson / IllustrisTNG Collaboration

p. 149: Jana Žďárská

p. 167: NASA / Chandra X-ray Center, D. Clowe, M. Markevitch

p. 177: Museums Victoria

p. 192: NASA / ESA / A. Riess (Space Telescope Science Institute / Johns Hopkins University) / S. Rodney (Johns Hopkins University)

p. 206: Wil Tirion—Uranography & Graphic Design

p. 212: ESA / Planck Collaboration

p. 226: XENON Collaboration

p. 239: Reidar Hahn, Fermilab

p. 254: NASA

p. 268: Courtesy Pieter van Dokkum

p. 284: NASA

p. 287: Press Office Karlsruhe Institute of Technology

p. 304: CERN

p. 311: ESA / C. Carreau

Index

Page numbers in *italics* refer to illustrations.